Territorial Ecology and Socio-ecological Transition

To Céline who supports me, day after day

Smart Innovation Set

coordinated by
Dimitri Uzunidis

Volume 34

Territorial Ecology and Socio-ecological Transition

Nicolas Buclet

iSTE

WILEY

First published 2021 in Great Britain and the United States by ISTE Ltd and John Wiley & Sons, Inc.

ISTE Ltd
27-37 St George's Road
London SW19 4EU
UK

www.iste.co.uk

John Wiley & Sons, Inc.
111 River Street
Hoboken, NJ 07030
USA

www.wiley.com

Library of Congress Control Number: 2020950525

British Library Cataloguing-in-Publication Data
A CIP record for this book is available from the British Library
ISBN 978-1-78630-546-6

MIX
Paper from responsible sources
FSC® C013604

Contents

Acknowledgments

This work has benefited from the support of the French state and in particular, within the framework of the following research projects:

– Projet Trajectories, within the framework of IDEX Université Grenoble Alpes, as part of the "Investments for the Future" program bearing the reference ANR-15-IDEX-02;

– CAP-Bioterproject, funded by ADEME (*Agence de l'Environnement et de la Maîtrise de l'Energie*, French Environment and Energy Management Agency), REACCTIF program;

– BOAT project, funded by ADEME (*Agence de l'Environnement et de la Maîtrise de l'Energie,* French Environment and Energy Management Agency), GRAINE program.

This publication was also made possible thanks to the exceptional working environment that I enjoy, whether at the PACTE laboratory and the IUGA (*Institut d'Urbanisme et de Géographie Alpine*, French Institute of Urban Planning and Alpine Geography), whose members and management I would like to thank, or with the colleagues involved in the *Trajectories* project, a particularly enriching interdisciplinary project on the trajectories of socio-ecosystems. In this respect, I would like to thank Sandrine Anquetin and Sandra Lavorel with whom I had the honor of co-facilitating this project. I would also like to thank Juliette Cerceau, who has helped me a lot in the CAP-Bioter project, which she has carried on with despite administrative adversity.

My thanks also go to the colleagues with whom I have been able to have rich and friendly exchanges with in recent years, whether in Grenoble or elsewhere. In no particular order, I would like to thank Sabine Barles, Gilles Billen, Muriel Maillefert, Sabrina Brullot, Philippe Lescoat, Sophie Madelrieux, Pauline Marty, Pierre-Antoine Landel, Myriam Donsimoni, Magali Talandier, Kirsten Koop, Gilles Debizet, Bernard Pecqueur, Claude Janin, Dany Lapostolle, Pierre-Yves Longaretti and Jean-Yves Courtonne. I have surely forgotten some colleagues and I apologize for this in advance.

Finally, I thank my family Céline and the children, whom I love above all else and whom I am so proud of. I, for one, am not so proud of the world we are leaving them, even though with books like this I try to do my part.

Introduction

For several years now, with a few colleagues involved, especially the participants in the Aussois adventure[1], we have typically presented territorial ecology as an emerging interdisciplinary field. We do not know until when the "emerging" qualifier will continue to make sense. It is certain that, despite some important research and wonderful achievements, territorial ecology is still a niche approach, whose methodology needs to be consolidated and must convince a large number of colleagues of its relevance, and even of its necessity. In France in particular, interdisciplinary work is complicated. Recruitment and evaluation procedures in the academic world are still marked by clearly defined disciplines. Interdisciplinarity transcends traditional boundaries, but it is here, we believe, that novelties are found, that well-established habits are shaken up and that certainties are questioned. In this sense, territorial ecology participates in this movement of questioning the compartmentalization of disciplines in order to elaborate innovative approaches and methods in the face of the socio-ecological problems.

This book has several objectives. The first is to situate territorial ecology, a niche interdisciplinary field, within a broader disciplinary family, that of *Ecological Economics*. In particular, it is important to establish proximities with other fields, be it Viennese *social ecology*, urban ecology or industrial ecology. 10 years ago, I published a book

1 See Buclet (2015), a collective work from a CNRS thematic school on territorial ecology, held in Aussois in 2013.

entitled *Industrial and Territorial Ecology*. Why this shift toward just territorial ecology with the abandonment of the industrial reference? The purpose of Chapter 1 is to educate the reader about these different aspects, as well as the relationship of territorial ecology with the concepts established by national and international institutions. The nascent territorial ecology was quite easily situated in the field of sustainable development. Things became more complicated, due to a certain dynamism from the institutions. What about territorial ecology in the face of the circular economy, the sustainable bioeconomy or the ecological transition? In this chapter, we would like to pay particular attention to an issue: that it is today the national and international institutions that are setting the conceptual agenda on which a large number of researchers are working. As legitimate as it seems to us that these institutions should share their political agenda and their priorities with the academic world, it is distressing to us that many researchers traipse around the concepts put forward by the institutions. This observation is not reassuring with regard to the state of a part of the research world, burdened by the need to respond to orders and obtain funding, rather than to think independently. To give an example, if the circular economy was initially a concept forged by researchers, it is clear that the meaning given to it today is not at all the same. This would in no way be a problem if it were not for the fact that researchers take up this new meaning and make it their field of research without questioning the consequences of this new meaning. This is why, in this chapter, even though we aim, with territorial ecology, to produce a way of thinking that can be useful to stakeholders, we wish to position ourselves in relation to what sometimes appears to us as a succession of new semantic arrangements that do not necessarily guarantee a renewal of thought and action.

The second objective of this book is to establish, especially for non-specialists as specialists might be disappointed by this too general overview, an inventory of the methods mobilizing the notion of metabolism. Metabolism, whether urban, industrial or territorial, is in fact the backbone from which to think about the analyzed system. The methods are diverse and depend largely on both the objectives sought and the disciplines mobilized, from biogeochemists to historians, urban planners and engineers. This inventory allows us to analyze what can be expected from each of the methods developed and the limits of these methods. More specifically, this allows me to distance myself from these works, most of which are remarkable, and open up a complementary path marked both by my own educational path as an urban planner and, earlier, as an economist, and by more operational goals.

The third objective, which forms the heart of our intentions, consists of proposing a method that will help to carry out an analysis of territorial ecology. This proposal is likely to form the backbone of territorial ecology, while being complementary to the other methods introduced in Chapter 2. It is based on the idea of giving more space to a qualitative analysis of the circulation of flows, in order to understand not only what circulates, but also why and how it circulates. In other words, in addition to a formal presentation of a territory's metabolism, broken down into subsystems that represent the various wealth-creating activities, we detail the different ways of analyzing the intentions behind the flows. As we shall see, this implies that we are interested not only in the circulation of flows of materials, water and energy, but also in the circulation of know-how, the influences that certain actors have on others, and the mobilization of intangible resources, such as heritage.

Thus, Chapter 3 of this book is based on the third chapter of the collective essay on territorial ecology applied to the municipality of Aussois in Savoie (Buclet 2015a). It details the method and removes certain inaccuracies and vagueness that still remained in the previous book. In Aussois, we sketched out a new method based on the analysis of the wealth-creating activities of a territory. Here, we present some of the advances obtained with this method, based on the questions that have arisen in recent years, in the context of various research projects to which I have contributed. As much as possible, this chapter guides the reader who may be considering using this method to analyze a territory. It presents how to produce a representation of the territory by wealth-creating subsystems, and the complementary analysis tools that make it possible to analyze the relationships between stakeholders that underlie these circulating flows, whether it be an analysis of the power games between them, the disparity that may emerge within the territory as to the orientations to be given to a project, or the motivations of one or the other.

Once the foundations of the method set, we wanted to go further and deal, in particular, with the question of co-evolution between human societies and their environment. Indeed, the main criticism that can be made about the method is that undoubtedly, it does not sufficiently take into account the interactions between nature and society. If we consider that a territory is also a socio-ecosystem, that it is not only the preferred scale for analyzing relations to our environment, but also the preferred scale for action at the interface between society and nature, then it is important to develop a grid for analyzing these interrelationships. In order to do this, we rely on the concept of capabilities, which we owe to the Indian economist Amartya Sen, as well as on the notion of territorial resources. Chapter 4 starts from these notions and concepts and revisits them in order to better articulate them with the issues of territorial ecology. We thus move from capability to territorial capability, which consists of a collective capability located at the scale of a territory and whose sustainability depends on socio-ecological balances. We also move from the notion of territorial resources to that of socio-ecological resources, in order to better highlight the fact that it is a co-production between nature and society. On the strength of these revisited notions, we present the stakes of a territory subjected to the socio-ecological constraints that it faces in this chapter. Indeed, we cannot ignore the need to confront our work with the climatic emergency, as well as with all of the ecological imbalances we are facing. Our proposal in this chapter is to articulate the metabolic approach with the notion of territorial capability. We believe that this articulation is likely to help territories to project themselves in the years to come with the aim of reconciling territorial development and environmental issues.

Elaborating a territorial project using territorial ecology, as with any other method, may however be insufficient, given the systemic complexity of sustainability issues. In particular, we cannot overlook the fact that the risk of collapse is the product of the dominant economic system. Thus, in Chapter 5, we will rely on the economics of conventions, and in particular, the notion of conventional regimes, in order to situate sustainability issues. In this chapter, we argue that it is misleading to think about sustainability without radically questioning the reference points that are at the origin of the very problems we intend to solve. This chapter, therefore, aims to present a certain number of principles on which to base an alternative to the dominant conventional regime. I will show that many local initiatives are proving to be bearers of these principles that transform the benchmarks and values that form the basis of a society. In conclusion, I invite the institutions to take into account this transformative, potential source of societal innovation and to use it as a basis for action at the territorial level, and to generate dynamics favorable to the emergence of a sustainable society. In doing so, I hope that this book will convince territorial stakeholders, and even national institutions concerned by the issue of sustainability, of the need to initiate debates around projects about living together, and to base these projects on principles likely to orient individual and collective actions in a sustainable direction.

1

Positioning Territorial Ecology in the Scientific and Institutional Context

Science within a given paradigm, recognized for its validity by a community of researchers, advances by successive additions. Territorial ecology, having been developed under the impetus of French-speaking researchers for about 15 years, is the result of an intellectual evolution considering environmental issues using a systemic approach. This interdisciplinary field has its roots in various lines of thought, most of which go beyond disciplinary boundaries, or simply do not take them into account, whether it be Ecological Economics, Industrial Ecology or Social Ecology. It is also in line with work aimed at understanding nature-society relations from a systemic approach, following the example of the work of the *Club of Rome and the Meadows Report* (1972).

However, it is not only the scientific community that will be considered. In the field of sustainability, scientific knowledge is expressed in many ways with the interplay of institutions. Of course, both local and global institutions feed on what the world of research can bring them, but we sometimes forget how much the reverse is true. In particular, some of the notions mobilized by the scientific community come from the discourse of international institutions (the UN for sustainable development), from experiments carried out by stakeholders in the field (Rob Hopkins for ecological transition), or from the initiative of national institutions (for example, the circular economy in France, a notion that emerged during the *Grenelle de l'Environnement*[1] although its origin is initially scientific).

Our objective is to not only understand which line of thought territorial ecology stems from – research has already been seamlessly carried out on this by Sabine Barles (see, for example, Barles 2010), which we will return to shortly – but to also grasp what territorial ecology is not, given the multiplication of notions that institutional and operational stakeholders grasp. Indeed, the field we are interested in is frequently disturbed by the semantic evolutions practiced by these stakeholders. Thus, the French institutional discourse changed overnight, without taking into account the possible consequences of this change, from industrial ecology to circular economy. Better still, the discourse has resized the limits of industrial ecology without any reflection on what this could mean (see Buclet 2015a). In the same way, the transition from a sustainable development policy to an ecological transition policy was made within French institutions without any particular discussion on the meaning of this change. The name of the ministry in charge of these issues has shifted from one to the other without any explanation of what this change entailed.

What does territorial ecology represent in this context, influenced by both research and political considerations? This is what we will present in this chapter, starting with the research aspect.

1.1. Positioning of territorial ecology from a scientific point of view

Territorial ecology refers to the idea of linking ecological issues with territorial issues. It is in line with work aimed at reconciling environmental sciences and human and social sciences, in order to couple them with the objective of better understanding what we call socio-ecosystems (Ostrom 2009). This interdisciplinary field contributes

1 The *Grenelle de l'Environnemen* is a collective work initiated in 2007 by the French government to determine ambitious advances in environmental actions. Many legislative texts that have subsequently been adopted have resulted from this.

to bringing together the challenges of human societies as understood by the human and social sciences and "the development of scientific ecology and in particular, the ecosystem theory" (Barles 2004, p. 75). The approach is intended to be systemic, following the example of the study of ecosystems, and based on questions related to thermodynamics (Georgescu-Roegen 1971). As Eugen Odum (1953, cited by Barles 2004, p. 76) indicates, "living organisms, ecosystems and the biosphere as a whole possess the essential thermodynamic characteristic of being capable of creating and maintaining a state of internal order, or low entropy."

This approach, coupled with the observation of the finiteness of the planet, leads to a consideration of the limits of an industrialized society, voracious in natural resources. While the discovery of the energy power of first coal, then oil, seemed to have brought humanity out of a situation of chronic scarcity, given the needs expressed, and even ushered in the era of unlimited abundance[2], the ecological and entropic cost of this power has been critical for the balance of the biosphere for several years now. This period of supposed abundance will only last for a few decades, which is very little compared to the scale of humanity.

1.1.1. *Urban ecology*

Considering systemic issues is not limited to the global level. As Sabine Barles (2004) reminds us, urban systems are also concerned, in that urban lifestyles are rightly considered particularly resource-intensive. This author quotes Lewis Mumford and his

2 This vision is highly questionable and is also contested by authors such as Marshall Sahlins and his book *Stone Age Economics*, published in 1972, and Mahjid Rahnema and his book *Quand la misère chasse la pauvreté*, published in 2003. These authors highlight, through their view of other civilizations, the apparent paradox of an industrialized society obsessed by the need to constantly find new resources to meet the needs of incessant growth, while the populations of these other civilizations, which we would describe as miserable according to our standards of living, seem to enjoy a sense of abundance or, at the very least, do not seem to know misery or poverty.

criticisms of megalopolitan gigantism as early as the 1960s. However, above all, Abel Wolman should be remembered here. The article he published in 1965 linked the city to the concept of metabolism for the first time and made this author, alongside Eugene Odum (Barles 2004), one of the founders of urban ecology. He was interested in the material balance (food, water, energy, waste, etc.) of an American city, both the inputs and outputs, and thus highlighted the extent of what he called metabolic problems.

In practice, these problems are reflected in the existence of poor sanitation infrastructures and major air pollution problems (Barles 2004, p. 80). The solution, according to Wolman, is technical and involves investment in infrastructure capable of resolving these metabolic imbalances. The link with ecology still remains rather tenuous, but opens up important perspectives in terms of both theoretical and operational research around the concept of metabolism.

These works, as well as those of Eugene Odum (1963), and then ecologist Paul Duvigneaud (1974), consecrate the idea of the city being studied as an ecosystem. Like Lewis Mumford, Paul Duvigneaud denounced the impact of the gigantism and immoderation of cities, driven by commercial processes (Duvigneaud 1974, p. 245).

Thus, beyond global ecological issues, which are becoming more topical with each passing day, the impacts of the most anthropized ecosystems, in this case cities, are increasing in interest. According to Sabine Barles, the success of this approach is due to the fact that it is based on a systemic approach (Barles 2004, p. 85) familiar to many researchers.

At the price of an analogy with the living world, the city is considered to only be able to live by drawing important resources from its environment, which led Eugene Odum to say that the urban ecosystem is a parasitic ecosystem (Odum 1975). This idea of parasitism was adopted by other authors and established the idea of a city's total dependence on other ecosystems. This dependence, which is inherent in the urban environment, seems to increase with the size of cities and poses a problem, because the increase of this environment is detrimental to other ecosystems. According to Eugene Odum, the challenge is then to transform the city, reducing its status as a parasite in order to increase possible symbioses with the environment.

This analogy has, as Sabine Barles (2004) points out, been very successful. The idea of comparing the city to an ecosystem that should be studied in the same way as natural ecosystems is taken up by several researchers. Its degree of complexity is certainly considered to be higher, but this should not be an obstacle to avoid. The city rightly enters the field of systems science and can be considered using a technical approach that involves identifying and quantifying the flows that circulate within the system and between it and its external environment.

This approach is attractive, but there are some limitations, as pointed out by Sabine Barles (2004, p. 86). The first, and in our view the most important, is that of the place of humans in the ecosystem under consideration. Humans are not considered as social beings, whose behaviors depend on principles, values, desires and feelings that motivate their actions within the urban ecosystem and thus explain the flows. According to this approach to urban ecology, humans are considered as functional beings, according to metabolic needs classified in very generic categories (living, working, circulating, recreating, etc.). Reducing humans to this mechanistic vision raises serious questions and, above all, reduces the scope of what urban ecology can provide.

Nevertheless, the approach of considering the city as an ecosystem first, and then as a living organism[3], will establish a method that analyzes the metabolism of the organisms that cities represent and will encourage more studies on their ecological footprint and on the load capacities of the systems studied.

3 The Earth is perceived as a living organism (see James Lovelock's Gaia hypothesis). In the same way, we will increasingly rely on the idea that highly anthropized environments, such as cities, can be analyzed as living organisms.

1.1.2. *Industrial ecology*

Another field of research into the relationship between human societies and their environment emerges: industrial ecology. If urban ecology is built from a strong analogy between cities and living systems, industrial ecology limits itself to considering this analogy as a source of inspiration. Moreover, beyond some pioneering work (such as the work of a group of Belgian researchers; Billen *et al.* 1983), industrial ecology will very quickly distinguish itself from urban ecology by its marked operational purpose.

This can be explained by the fact that, beyond previous work, such as that mentioned above on the Belgian ecosystem, and in the opinion of Suren Erkman (2004) in particular, the founding article, not on industrial ecology, but on the craze for this oxymoron, was written by Robert Frosch and Nicholas Gallopoulos of General Motors in 1989 (Frosch and Gallopoulos 1989). These two authors, vice-president of research and head of engine research at General Motors, respectively, consider that:

> the traditional model of industrial activity – in which individual manufacturing processes take in raw materials and generate products to be sold, plus waste to be disposed of – should be transformed into a more integrated model: an industrial ecosystem. (Frosch and Gallopoulos 1989, p. 144)

Later they indicate that the industrial ecosystem should function in analogy with a natural ecosystem (ibidem). They write further:

> An ideal industrial ecosystem may never be attained in practice, but both manufacturers and consumers must change their habits to approach it more closely if the industrialized world is to maintain its standard of living – and the developing nations are to raise theirs to a similar level – without adversely affecting the environment. (Frosch and Gallopoulos 1989, p. 144)

It can be said that this article is a kind of manifesto of the principles of industrial ecology, to make society evolve toward more sustainability, without calling into question the dominant economic system, on the contrary destined to be generalized to developing countries. It is far from both the main aim of urban ecology, which aims to explain the functioning of cities through a naturalistic and systemic approach, and its critique of modernity. Engineering sciences are at the heart of industrial ecology, with a more operational vision and an optimizing purpose. The editorial of the first issue of the *Journal of Industrial Ecology*, in January 1997, sets the tone:

> A systems-oriented approach that integrates economic and environmental phenomena is crucial. Industrial Ecology [...] focuses on the role of industry in reducing environmental burdens throughout the product life cycle, from the extraction of raw materials, to the production of goods, and to the management of the resulting wastes. (Lifset 1997, p. 1)

The first issues of the same journal are devoted to this guideline with the following titles: "Slow progress in Ecolabeling: Technical or Institutional Impediments"; "The View from Australia"; "The Virtual Ecology of Industry"; "The Industrial Ecology of Lead and Electric Vehicles"; "Life-Cycle Assessment: Constraints on Moving from Inventory to Impact Assessment"; "The Relationship between Cleaner Production and Industrial Ecology"; "Industrial Ecology in Practice: The Evolution of Interdependence at Kalundborg"; "The Life-Cycle of Chlorine, Part I: Chlorine Production and the Chlorine-Mercury Connection"; "Chlorine in the Netherlands, Part I, An Overview"; "Going Organic: Converting Patagonia's Cotton Product Line" and "Recent Advances in Design for Environment at Motorola". In this first issue, the issues of pollutants and clean production are highlighted, as well as case studies, with a view to reducing pollutant impact and optimizing the use of resources.

This work has two focuses: methodological work, aimed at increasing the diagnostic capacities of productive processes, via industrial metabolism or life cycle analysis and monographic work on industrial ecology experiments around the world. The technique of *success stories* will be mobilized, with the Kalundborg symbiosis in Denmark as an example. Long before the publication of the first issue of the *Journal of Industrial Ecology*, several publications highlight the success of this industrial symbiosis[4]. Nicholas Gertler's thesis, defended in 1995 at the Massachusetts Institute of Technology (MIT), provides a stable basis for the success of Kalundborg through a detailed analysis of the reasons for its success. Kalundborg presents itself as a paradox. It is a small town of about 20,000 inhabitants with an industrial port, whose activities are centered around a thermal power plant for electricity production and an oil refinery. In other words, the typical example of the success of industrial ecology, which is supposed to contribute significantly to increasing the sustainability of our industrial society, is based on the massive use of non-renewable resources with a strong impact on climate change.

Kalundborg's results, as quantified by Nicholas Gertler, are interesting. Significant savings in water, materials and energy contribute to making Kalundborg a place of industrial tourism for anyone interested in implementing industrial ecology. The analysis even shows that the return on investment of operations to set up eco-industrial synergies is quite short, between three and five years (Gertler 1995). This raises the question of the implementation of comparable projects to the Kalundborg, elsewhere in the world, forgetting certain factors of success in the Danish case, notably the lack of willingness to make industrial ecology (a term that did not exist in 1961, when the municipality began to develop initiatives) and a slow progression in the implementation of synergies (few but decisive synergies during the first 30 years, followed by a singular acceleration from the 1990s[5]). In other words, after discovering Kalundborg and its spectacular results in terms of flow looping, researchers and practitioners of industrial ecology set themselves the objective of understanding what conditions are needed to establish other

4 Among the first publications on this theme is Nicolette Perry's book (1990).

5 Available at: https://www.lampindex.com/2011/10/the-kalundborg-symbiosis/ (accessed January 3, 2020).

eco-industrial symbioses throughout the world and initiate similar dynamics.

It is also in this sense that industrial ecology differs from urban ecology. The analogy pursued in practice (and sometimes even in research) becomes just as much, if not more, a sought-after analogy between Kalundborg (and other less striking *success stories*) and other industrial zones, as between industrial zones and natural ecosystems. John Ehrenfeld and Marian Chertow even speak of the Kalundborg legacy (2002) as a structuring symbiosis, a model highlighting not only the importance of co-operation between entities, but also the possible financial, as well as environmental, profitability of such symbioses.

From then on, many works strived to discover the key to the success of an industrial symbiosis in order to multiply successful experiences. Mainly driven by the engineering sciences, industrial ecology, like Nicholas Gertler's work in Kalundborg, is "discovering" the importance of the "human factor", the undocile variable that disrupts the best planned operations and that has been ignored for so long in favor of optimization on paper.

1.1.3. *From industrial ecology to industrial and territorial ecology*

Taking into account this "human factor" will immediately feed more reflections in the francophone community. The founding moment of a French-speaking scientific community was the organization of an international conference on the subject at the Université de Technologie de Troyes (UTT) in September 1999. Under the joint impetus of Dominique Bourg and Suren Erkman, co-organizers of this conference, a small community of researchers and practitioners was formed between engineering and social sciences, with the UTT as its epicenter. The challenge is to reflect, notably through fieldwork (Club d'écologie industrielle de l'Aube, Ecopal at Grande Synthe, then other fields elsewhere in France ...), on the institutional means of triggering symbiotic dynamics between industrial stakeholders. In conjunction with local public authorities interested in the approach, discussions are underway to identify flows of energy, water or materials likely to initiate synergies between players. Through research programs funded by the Ademe or ANR (COMETHE (Design of Methodological and Evaluation Tools for Industrial Ecology), CONFLUENT (Knowledge of Urban Flows, Environmental Footprints and Sustainable Governance), ARPEGE (Workshop of Prospective Reflection on Industrial Ecology) projects), the French-speaking community is structuring and deepening the organizational and institutional issues related to eco-industrial symbioses[6]. Under the influence of practitioners such as Benoît Duret (2006), for example, this community is beginning to talk about industrial and territorial ecology, which is more meaningful for local public decision-makers in charge of these approaches. Benoît Duret, after an analysis of a significant number of initiatives in Europe and North America, made 10 recommendations for the "development of industrial and territorial ecology" (2006, p. 38). These recommendations are as follows: develop an ecosystem approach to territories; explore trends related to the networked innovation economy; increase the support of local authorities to cooperation networks; better capitalize and share experiences of industrial ecology; inform, sensitize and train state institutions and local decision-makers; integrate industrial ecology into existing sustainable development practices; articulate industrial ecology with territorial foresight; perpetuate approaches through the creation of dedicated animation structures; build and make available, information on the physical flows of the economy[7]; and associate industrial ecology strategies with those of the social and solidarity economy. As we can see, we are quite far from an engineering approach and are more on the scale of social issues, inspired by social science concerns. The territory is mentioned, both in terms of the geographical scale and in terms of cooperation between the stakeholders concerned, and brings a specifically French-speaking slant.

6 See in particular the PhD by Sabrina Brullot (2009), which aims to establish a decision-making process that is favorable to the emergence of a territorial dynamic around industrial ecology.

7 In particular, the aim is to inform the flows, in order to have a metabolic approach to the industrial system under consideration.

1.1.4. *The social ecology of Vienna*

French-speaking research in industrial and territorial ecology is resolutely oriented towards operational research on the conditions for the emergence of industrial symbioses. Austrian research, under the impetus of Marina Fischer-Kowalski and Helmut Haberl, takes a rather different direction, still based on the method of metabolism, the common point between all of these approaches. Marina Fischer-Kowalski's choice to move away from *industrial ecology* towards *social ecology* may surprise on several counts. First of all, because Vienna's *social ecology* has nothing to do with the *social ecology* that has developed throughout the world, particularly in the United States, and was founded by the anarchist activist Murray Bookchin. Second, the *Institute for Social Ecology* in Vienna specializes in the analysis of material flows and the development of quantitative methodologies, for a long time neglecting the contribution of the social sciences, or at least placing it largely in the background.

Social Ecology in the world refers very explicitly to Murray Bookchin. Both the *Institute for Social Ecology* in Vermont and the *Transnational Institute for Social Ecology,* located in Athens, Greece, take a radically critical stance towards society and the dominant economic system[8]. This posture is absent in Vienna, however. Researchers at the Austrian institute pursue a goal of knowledge rather than operationality. While the quantitative option is favored, it is original, with a significant contribution from ecology and a historical approach to the cities and territories analyzed, in order to describe the present, as well as the past of the fields of study. The idea of co-evolution, and the analysis in time and space of this co-evolution between nature and society, is affirmed as the Institute's main idea. This guideline has been reinforced in recent years, with a growing weight of social science researchers in the team, so that the relations

8 "Social Ecology advocates a reconstructive and transformative outlook on social and environmental issues, and promotes directly democratic, confederal politics. Social Ecology envisions a moral economy that moves beyond scarcity and hierarchy, towards a world that reharmonizes human communities with the natural world, while celebrating diversity, creativity and freedom." Available at: http://social-ecology.org/wp/about/about-the-ise/ (accessed January 5, 2020).

between nature and society, beyond the material aspect, can be worked on further. In this sense, we can say that the Viennese *social ecology* and territorial ecology presented in this book tend to come closer together.

1.1.5. *Toward a territorial ecology*

Like all the fields of research presented so far, territorial ecology is based on the idea that the analysis of the metabolism of human societies provides structure. First amongst other researchers, Sabine Barles spoke explicitly about territorial ecology, and many works were developed which, afterwards, were considered to be fully in this field. In particular, this is the case for work that could be considered close to Viennese *social ecology*. It aims to develop methods for analyzing the footprint of cities, in the past as well as in the present, based on a better understanding of the mobilization of flows of materials, water and energy. The special issue published in *Regional Environmental Change* in 2012 (volume 12, number 2), following a workshop organized in Paris in 2009, at the initiative of Gilles Billen and Josette Garnier, is emblematic of this approach. Entitled *History of Urban Environmental Imprint*, this special issue features an article on the metabolic weight of Brussels on the surrounding countryside in the Middle Ages, on the way London was nourished in the Middle Ages, on the use of resources in Vienna in the 19th century (work identified as being in *social ecology*), on the energy consumption of Paris from the 18th century to the present day and on the food consumption of Paris or New York for the same period of time, on the consumption of non-ferrous metals in Paris from the 19th century to the present day and on the water consumption of Athens, Barcelona and New York in the past and present. We talk about food, energy and water footprints and focus on the material consequences of the way cities function. What distinguishes this work more particularly from urban and industrial ecology is the remarkable effort used to spatialize the phenomena studied, to map the interactions between cities and "their hinterland" and in particular, the weight of the former on their immediate environment.

It is possible to distinguish what potentially falls within the field of territorial ecology (see above) from what explicitly claims to be territorial ecology. For it is only in the French-speaking framework, following the example of industrial and territorial ecology, that authors speak of territorial ecology and do so as an interdisciplinary field. Initially, territorial ecology is evoked as an approach resembling industrial ecology, "aiming at optimizing the use of resources and better managing the associated impacts" (Duret *et al.* 2007, p. 73), except for one detail: it does not apply specifically to industry but to a whole territory.

This distinction is in itself not very satisfactory, and with this first reference it is difficult to perceive the difference between industrial and territorial ecology on one hand and territorial ecology on the other hand. In the same way, as Sabine Barles wrote quite early on, referring to one of her works:

> the approach is thus inscribed in the field of territorial ecology, a declination of industrial ecology that allows taking into account the multiple interactions between urban activities and the biosphere, and is able to contribute to the governance of material flows. (Barles 2007, p. 79)

At first glance, territorial ecology would only be a nuance of industrial ecology. But this subtle nuance will gradually gain in importance:

> Territorial ecology constitutes an emerging interdisciplinary field of research in France, which is similar to urban ecology (when it is based on ecosystem theory) and industrial ecology (when it is devoted to industrial societies) without being confused with them. (Barles *et al.* 2011, p. 16)

A double filiation is thus assumed:

> Its main objective is to better understand the nature and modalities of interactions between societies and nature,

> the main concrete manifestation of which lies in the energy and material exchanges that take place within and between anthroposystems. (ibidem)

Territorial ecology thus has the purpose of knowledge more than operationalization, which tends to distinguish it from industrial ecology, which is dedicated to work on industrial ecosystems.

The differences, if not in method then at least in intention, will tend to increase over the years, without denying the link with urban ecology on one hand and industrial ecology on the other hand. It is therefore the intention that varies. It is from this variation, which is above all an evolution, that the work in territorial ecology will progressively move away from the work in industrial ecology.

The points that we feel are appropriate to highlight at this stage are the following: the orientations of the scientific work and the posture with regard to the challenges of sustainable development.

On the first point, we have already underlined a less directly finalized orientation of the work in territorial ecology than in industrial ecology. The latter is largely driven by engineering sciences, aims to optimize processes and identifies sources of improvement with regard to the potential identified using metabolism analysis or life cycle analysis. This trend can be seen in most of the scientific literature on industrial ecology, particularly in the published issues of the *Journal of Industrial Ecology*. Territorial ecology aims to understand the functioning of human societies through their materiality. In fact, the analogy between natural and anthropogenic ecosystems, at the heart of industrial ecology, is a question that is not very present in territorial ecology. It is a probable consequence of this lack of operational finality, of the need for a utopia to be pursued that is found in industrial ecology (see Allenby *et al.* 1994). In territorial ecology, action is not ignored, but it requires "the development of knowledge that is not directly finalized" beforehand (Buclet 2015b, p. 19). Consequently, what is important, when faced with the objectives of flow closing loop and a less open metabolism, is not to be inspired by nature, but to work towards making these objectives meaningful, given the stakes of sustainable development, by taking greater account of ecosystem balances. The understanding of society-nature relations at the scale of the territory is at the heart of territorial ecology, which implies taking the territory as the object of study in its own right:

> Through the production of knowledge on this territory, it participates in its definition and its material, organizational and identity structuring. (Buclet 2015b, p. 19)

Work in territorial ecology is more geographically situated and gives a place of choice to the human and social sciences.

The stance of territorial ecology with regard to sustainable development, constitutes the other aspect likely to better differentiate this interdisciplinary field from industrial ecology. Some writings tend to indicate that a change of paradigm would be necessary (Ehrenfeld 2000), or in any case a reflection on the question of development and its relation to economic growth (Druckman and Jackson 2016). However, the positioning of most of the work in industrial ecology is within the framework of a sustainable development, which could be achieved without calling the main foundations of the dominant economic system into question, with the help of progress or even technical and/or organizational innovations; not that the authors in industrial ecology are opposed to questioning the system but, obviously, this question is only marginally considered in the publications directly.

Although this questioning is not inherent to territorial ecology, it is more relevant to it in that: (1) territorial ecology does not specifically concern industrialized society, and the work carried out is not only confined to the analysis of the dominant conventional regime[9], but

9 "A conventional regime corresponds to a moment in time and space during which a given human society conforms to a certain number of principles that guide the behaviour of individuals in that society" (Buclet 2011a, p. 66). The dominant conventional regime is the regime based largely on the principles of a capitalist, market, industrial and globalized society (Giddens 1990). This will be discussed at length in Chapter 5.

also of other regimes that have structured societies in the past or, more rarely, in the present. This *a priori* presupposes less the way in which a human society thinks. (2) Although technology is very much present in work on territorial ecology, it is not the instrument of sustainability at the service of which to think and act. (3) The question of sustainable development is not only environmental (and cannot be stupidly reduced to a crossroads between the economic sphere, the social sphere and the environmental sphere, as is too often presented). It is closely linked, as we have wished to highlight in recent years (see Buclet 2015b), to the question of how humans live within a given territory and what means they have at their disposal. It also questions the collective capacity to reconcile the challenges of limited resources with the challenges of living together, and the quality of individual and collective life. In this sense, many researchers who claim to be concerned with territorial ecology allow themselves the freedom to reflect on alternatives to the dominant conventional regime, and assume this posture. In Chapter 5 we will see that the question of innovation in territorial ecology is posed in terms of the transformation of the reference points that structure individual and, above all, collective action. A fundamental bridge is thus established between quantitative methods, the first of which is the analysis of territorial metabolism, and a critical stance questioning the foundations of a system in the throes of collapse, both from an ecological point of view and from the point of view of its social legitimacy, since it is becoming so difficult to follow the trajectory of protest movements of all kinds.

1.2. Positioning in the institutional context

These last considerations, which some may find very unscientific – as if science could be neutral – allow us to move onto the positioning of territorial ecology in the institutional context. Research is in fact influenced by the way in which institutions, local but above all national and international, deal with social issues. This is all the more true in recent decades through the mechanisms of programing research according to societal issues.

1.2.1. *The evolution of concepts around sustainability*

It seems necessary for us to come back to the evolution of the stakes and concepts mobilized by national, European and international institutions in relation to sustainability issues. Since the publication of the United Nations report in 1987, known worldwide as the Brundtland Report, institutions have wished to highlight sustainable development as a mobilizing concept. The advantage of this concept, as shown by several authors, notably Edwin Zaccaï (2002), is that it corresponds to both the setting of an agenda of a strong need to react to the alerts launched by the scientific world on the viability of our society within the biosphere, and to the belief in the capacity of the dominant conventional regime to face this need without fundamental questioning. If the double term of sustainable development has gradually become widespread, to the point of becoming an obligatory passage to legitimize any public or private action or project, other terms have appeared that clearly show the ambiguity of the use made of it. Thus, we speak of "sustainable growth", which would be "by definition economical in raw materials, low in pollution and not linked to financial imbalances" (Artus 2011). The two 2018 Nobel Prize laureates in economics were even rewarded for having "developed methods that respond to some of the most fundamental and pressing challenges of our time: combining the long-term sustainable growth of the world economy with the well-being of the planet's population."[10] Better still, with the adoption of Sustainable Development Goals, Goal 8 is entitled: "Promote sustained, shared and sustainable economic growth, full and productive employment and decent work for all."[11] With such assertions, there is no doubt that the shift from sustainable development to sustainable growth proves that all those who criticized sustainable development as a notion that does not question the behavior of the stakeholders of the dominant conventional regime are right, but this accompanies the movement (Ariès 2005). How can we still defend the idea that development is qualitative when growth is quantitative, even though both are intertwined in the discourse of

10 Available at: https://lexpansion.lexpress.fr/actualite-economique/la-croissance-durable-priorite-des-deux-nobel_2038940.html (accessed January 9, 2020).

11 Available at: https://www.un.org/sustainabledevelopment/fr/economic-growth/ (accessed January 9, 2020).

certain researchers and national and international institutions? The posture of authors (such as Gilbert Rist (2001), who considers sustainable development as the umpteenth way of making the concept of development bounce back) comes out, reinforced from these amalgams and stances that mitigate the question of sustainability.

In support of sustainable growth, we should also mention green growth, an even more restricted term, but one that is regularly promoted as a way to improve sustainable development. Thus, the OECD, a key institution for guiding thinking, published a document in May 2011, for the Heads of State and Ministers of more than 40 countries entitled the "Green Growth Strategy"[12]. The main question covered in this document is the following: "What is green growth and how can it help to ensure sustainable development?" In particular, it states that green growth is at the intersection of two challenges: "expanding economic opportunities for all in the context of a growing world population; and addressing environmental pressures that, if left unaddressed, could undermine our ability to take advantage of these opportunities"[13]. The paper goes on to explain that "Green growth is not a substitute for sustainable development. Rather, it is a practical and flexible approach to achieving concrete and measurable progress across all of its economic and environmental pillars, while taking full account of the social consequences of greening the growth dynamic of economies."[14] As it can be seen, these strong assertions do not call the foundations of the dominant conventional regime into question.

In spite of all of this, despite so many certainties displayed by institutions about what should be done, several elements have forced them to modify their discourse, or rather the semantics used. In particular, it is the litany of international scientific reports which, following the example of the International Panel on Climate Change (IPCC), conclude each time that their previous report was too optimistic in view of the worrying developments in the ecological situation. They

12 Available at: https://www.oecd.org/fr/croissanceverte/quest-cequelacroissanceverteetcommentpeut-elleaideraassurerundeveloppementdurable.htm (accessed January 9, 2020).

13 *Ibid.*

14 *Ibid.*

are also increasingly strong criticisms, shared by more and more stakeholders, against this ambiguous terminology. The public authorities in various countries have therefore considered it necessary to mobilize new, less contested terms. It is in this context that the concept of transition has emerged, thanks to certain scientific works (in particular, the work of Franck Geels on socio-technical transitions; Geels 2002). Just as there was talk of eco-development and a more radical critique of the dominant conventional regime (Sachs 1974) in the 1970s, sustainable development, which was more politically correct, replaced this concept, and ecological transition emerged as a change in the dominant way of mentioning sustainable development.

Initiated by Rob Hopkins and the City in Transition movement in 2006 in Totnes (the United Kingdom), the ecological transition triggers local dynamics, aimed at using permaculture principles to do without fossil resources (Hopkins 2010). The movement also advocates a different relationship with the economy, in particular a strong relocation of the economy, likely to strengthen the links, solidarity and cooperation between stakeholders in the same area. This is an ambitious movement, based on collective action at the local level, which will bring about changes in individual and collective behavior.

In a very short period of time, institutions have, in many circumstances, replaced references to sustainable development with the term "transition" coupled, depending on the situation and the preferences of the writers, with the terms adjective energy, socio-ecological, or ecological. Is it to be seen, as proclaimed by these institutional stakeholders, as a willingness to do better than with sustainable development. Following the example of Patrick Viveret[15], it is possible to doubt it. Thus, on August 17, 2015, the French government adopted law n°2015-992 relating to the energy transition for green growth, showing that the term "transition" does not in any way represent a desire for change, but a new transformation of growth that we would try to make green. Thus, it is better understood that a new concept, supposed to modify the scale of action (the local), or even question the belief in the superiority of the globalized market, is

15 Available at: https://www.liberation.fr/france/2014/08/26/le-mot-transition-est-un-terme-imprecis-et-insuffisant_1087459 (accessed January 9, 2020).

not rejected by national and international institutions, but on the contrary is absorbed, in order to replace the concept of sustainable development that is now considered obsolete. Let us hope that the transition will soon allow us, to paraphrase Giuseppe Tomasi di Lampedusa, to change everything so that nothing changes.

1.2.2. *From industrial ecology to circular economy*

Under the impetus of Dominique Bourg, already an initiator of industrial ecology in France as an operational and scientific concept, industrial ecology started to appear on the agenda of the French government. Several ministers in charge of the environment between 2002 and 2006 were asked by the philosopher to set up a plan to disseminate the concept at the operational level. The idea was to identify regulatory measures in order to encourage companies to take an interest in the approach and local authorities to develop eco-industrial symbiosis projects. As a result of what was presented earlier in this chapter, industrial ecology in practice in France essentially aims to identify potential synergies between flow-producing stakeholders, so that the waste of some becomes the resources of others. As a result, the stakeholders concerned, institutions such as companies and local authorities, come to think of industrial ecology as synonymous with eco-industrial symbioses.

This state of affairs was reinforced with the discussions of the *Grenelle de l'Environnement*, starting in the summer of 2007. Still under the impetus of Dominique Bourg, who was in great demand by successive governments, the term circular economy would gradually replace the term industrial ecology, with the latter being relegated to being just one of the strategies for implementing the circular economy. Thus, among the conclusions of the *Grenelle de l'Environnement*, it is noted that:

> regulations and practices should be changed to enable the promotion of industrial ecology (or circular economy). (Ministry of the Environment 2007, p. 43)

We see in passing that, at this stage, the two terms are still considered equivalent, but this cannot continue. Subsequently, a new institutional stance will take hold of the circular economy and develop its meaning. According to the ministry in charge of the dossier, this is "an economic model whose objective is to produce goods and services in a sustainable manner, limiting the consumption and waste of resources (raw materials, water, energy) and the production of waste." The aim is to break away from the linear economy model (extract, manufacture, consume, throw away, etc.), for a new "circular" economic model[16]. The same web page tells us that the circular economy is part of the field of the green economy. This new term increases confusion and another ministry, that of the economy, tells us that it brings together two types of activities: "traditional activities carried out with less polluting or less energy-consuming processes and eco-activities, whose purpose is to protect the environment or manage natural resources." These activities are part of a green growth objective: the aim is to promote sustainable and lasting development in the long term[17]. Considering the circular economy as an integral part of a green economy whose ambitions are nevertheless quite modest (since it is a question of green growth), makes us take the full measure of a permanent watering down of the discourse, as well as a real economization of sustainability issues.

Returning to the circular economy, which is now included in the aforementioned[18] law on the energy transition for green growth (August 18, 2015), the Ministry's website describes its areas of application. We find on the website that the industrial and territorial ecology is defined as a synergy and a mutuality between several economic actors of the flows of materials, energy, water, infrastructure, goods or services in order to optimize the use of resources in a territory[19]. Thus, in the

16 Available at: https://www.ecologique-solidaire.gouv.fr/leconomie-circulaire (accessed January 10, 2020).

17 Available at: https://www.economie.gouv.fr/cedef/economie-verte (accessed January 10, 2020).

18 But also integrated into an action plan of the European Union (Communication from the Commission to the European Parliament, the Council, the European Economic and Social Committee and the Committee of the Regions, COM (2015) 614 final, Brussels, December 2, 2015).

19 Available at: https://www.ecologique-solidaire.gouv.fr/leconomie-circulaire (accessed January 10, 2020).

French institutional panorama, industrial ecology disappears in two stages. After being replaced by the expression circular economy, it is reduced to eco-industrial symbioses, which is now called industrial and territorial ecology.

1.2.3. *A variation of circularity: the bioeconomy*

As the concept of circular economy is gaining ground, links with other concepts are gradually being forged. This is notably the case for the sustainable bioeconomy. The concept of the bioeconomy was developed by Nicholas Georgescu-Roegen (1971), who was very critical of the current economic system. According to the Romanian author, it is fundamental to take into account the principles of thermodynamics and especially the second one, which introduces the notion of entropy. Nicholas Georgescu-Roegen criticizes an economic system that generates an excessive entropy of the biosphere, i.e. an excessive disorganization of the living world, not only because of our withdrawals, but also because of our little consideration for ecosystem balances. The positioning of this author is quite close to the posture of authors who advocate a certain degrowth. However, following the example of the concept of eco-development, watered down through that of sustainable development, the term bioeconomy has been used in recent years in a completely different sense. Thus, the Ministry of Agriculture and Food, via its website, tells us that:

> "The bioeconomy is the economy of photosynthesis and more broadly of living things. It is based on the production and mobilization of biomass for an optimal valorization. It allows the maximum use of abundant, renewable and free energy: solar energy. The bioeconomy encompasses all activities of production and transformation of biomass, be it forestry, agriculture and aquaculture for food production, animal feed, biosourced materials and energy. It represents opportunities for our agricultural and forestry operations, it increases the competitiveness of our industries while providing sustainable solutions to the environmental and societal challenges we face today. The bioeconomy is part of the broader green economy – an economy that respects

> the environment and uses natural resources more efficiently."[20]

The Ministry's strategy mentions the first issue of the bioeconomy as: "making the products of the bioeconomy market realities" and refers to the European strategy (European Commission 2012) for which "the Europe 2020 strategy advocates developing the bioeconomy as a key element of green and smart growth in Europe."

In France, there is a need to "support the transition towards a high-performance, innovative and sustainable biosourced industry" (Ministry of Agriculture and Food 2016, p. 14) and in particular "to promote industrial ecology through synergies in the use of bioresources and production factors."[21] The injunctions highlight the optimization of bioresources valorization, with some inflections in relation to mass industrialization, since it is necessary to "develop innovative and flexible production systems that adapt to the resources used,"[22] but in fine, what appears from these strategic documents does not suggest a change in the way of weaving different links between nature and society. With the bioeconomy, nature will be required to supply increased production. While the Department regularly reminds us that this must be done in a way that respects the ecosystem balances of the agricultural and forestry environments, not once is any mention made of organic agriculture, agro-ecology or, not surprisingly, permaculture-type approaches.

1.2.4. *An effective consideration of the stakes*

This regular reinterpretation by national and international institutions of the concepts that are supposed to make us think about a serious change in our way of life, production and consumption, is not without a certain number of problems. For this reinterpretation, even more than a watering down, tends to give back legitimacy to the ways of doing things that have led to this situation of unsustainability.

---

20 Available at: https://agriculture.gouv.fr/la-bioeconomie-nouvelle-vision-du-vivant (accessed January 10, 2020).

21 *Ibid.*

22 *Ibid.*

Objectives considered as unavoidable necessities are regularly put forward, such as an increase in production or the massification of optimized processes. In addition, there are objectives inspired by industrial ecology, such as the optimization of industrialized processes through the looping of increased flows between processes, but which do not fundamentally change anything about the dominant principles that participate in the decision and action of the stakeholders. Rarely, if ever, are the consequences of this type of productivity logic taken into account. Of course, when reading these strategies carried by institutions that undoubtedly think that they are doing the right thing, it may seem exaggerated to multiply the criticisms. The productive system is called upon by public authorities to innovate in a direction that allows it to produce more with fewer resources, certainly one of the challenges of the coming decades. However, there is no guarantee, in all of these approaches, that we will actually succeed in consuming fewer resources overall. In particular, we must count on the rebound effect at several levels, individual or society as a whole (Herring 2006). The direct rebound effect consists of consuming more of something because of the fall in its price induced by increased efficiency in the production phase. To this effect an indirect rebound effect is added, linked to the fact that efficiency leads to lower prices and the consumer has an increased purchasing power that he or she will use to consume other goods. Finally, the general rebound effect is linked to a more global effect linked to the myriad of actions of both consumers and producers (Greene *et al.* 1999). Although no one can precisely measure the magnitude of these effects, particularly on a global scale, it is clear that the continuous increase in industrial productivity in no way translates to a reduction in the consumption of raw materials over the medium or long term. Rather, we are witnessing a continuous increase that tends make us think that, beyond population growth, each individual on the Earth tends to consume more and more resources. In other words, the entire world population consumes more and more each year, as we are unfortunately reminded by the determination of the "day of overcoming" which, from September 29 in 1999, has passed to July 29 in 2019.[23]

23 Available at: https://www.footprintnetwork.org/ (accessed January 12, 2020).

1.3. Conclusion

Territorial ecology finally remains on the sidelines of these institutional movements, notably because its historical bearers (Sabine Barles, Gilles Billen, as well as the author of this book) did not have the desire to convince and continue to convince national institutions of the absolute necessity to establish a sustainable development strategy based on territorial ecology. Would this be synonymous to saying that territorial ecology is only a sterile field of scientific knowledge? The answer is, in our opinion, quite different.

Starting from a less operational objective than industrial ecology, circular economy or ecological transition, territorial ecology aims to understand the relationships between the organization of human activities and the observable interactions with their natural environment. Understanding remains a prerequisite that we cannot sacrifice to the feeling of working for sustainability on a more or less operational basis. Yet this understanding can only be achieved in close cooperation with the stakeholders of the territories concerned by the studies carried out. Indeed, the interplay between material flows and the intentionality of stakeholders requires a dialogue capable of not only identifying and collecting data on the flows that circulate in and around a territory, but also of understanding the dynamics between all who make up a territory.

In the absence of national and international institutions to involve and convince, territorial ecology, as we will show in the following chapters, aims to bring elements of reflection to be shared with the stakeholders of the territories being analyzed. More than a strategy, this interdisciplinary field aims at producing knowledge that can take into account the socio-ecological stakes in the elaboration phase of territorial projects. The challenge of sustainability requires that innovation also be carried out at the level of territorial dynamics. We may not be able to do without technological innovations, but if these are not a part of the projects transforming society, particularly at the level of the experience of individuals, i.e. the territory, it will be difficult to expect them to lead us towards any kind of sustainability.

2

Main Methodologies Developed Around Urban and Territorial Metabolism: Contributions and Limitations

Territorial ecology is methodologically structured around the analysis of the metabolism of territories, whether urban or territorial. The method used, as indicated in Chapter 1, is generally shared with industrial ecology and social ecology. Nevertheless, there is a large variety of ways one can apply this analysis, according to the researchers' objectives. It seems important to us to present the different facets, what they bring and the interest that there may be in terms of complementarity between these works. Thus, in this chapter, we will first present the main principles of metabolism, before showing the different variations, and then highlighting the few limitations, to which it would be appropriate to provide answers.

2.1. The main principles of metabolism

The basis of metabolism is material flow analysis (MFA). This concerns the accounting of the physical flows within the predefined system, and also between this system and its external environment, as well as the material stocks within the system for the determined analysis period. The unit considered is generally a ton of material, and the total tons of all circulating materials are added up indiscriminately.

To avoid any ambiguity, it should be noted that this material can circulate in solid, liquid or gaseous form. On the other hand, water is generally not included in the MFA. Very large tonnages of water flows tend to make the other flows represented on the same diagram insignificant.

2.1.1. *Plurality of objectives and methods*

Several researchers worked collectively in the 1990s not to completely homogenize the methods used, but to establish a platform for exchanging information on the different methods developed (Bringezu and Moriguchi 2002). This collective work gave rise to the Eurostat guide (economy-wide material flow accounts and derived indicators) in 2001 and the OECD guide (measuring material flows and resource productivity) in 2008.

We will not go into the details of these different methods here, but we will resume discussion on the different implementation strategies identified by Stefan Bringezu and Yuishi Mizoguchi. Each method carries a specific objective. Thus, the first strategy is that of the "detoxification" of industrial metabolism, i.e. a strategy to better measure the impact of polluting substances on the analyzed environment. With a completely different objective, the two authors have identified a second strategy, called "dematerialization of metabolism", in order to reduce the impact of the analyzed system on the biosphere. It can be seen that the quantification exercise is very different in both cases since, in relation to the detoxification strategy, we focus on one or several substances, in order to follow their course, identify the level at which these substances are emitted and/or stored (in the event of a concentration threshold problem), over time[1]. This approach is characterized by the importance of determining when an d where the desired substances are dispersed in the environment, and therefore presupposes a coupled approach between biochemists,

1 An example is the work carried out on heavy metals in the Seine basin over periods of several decades (Meybeck *et al.* 2007) which, thanks to extensive historical study, has made it possible to identify possible contaminated sites due to excessive concentrations.

geographers and historians, which some researchers embody, like the biogeochemist Gilles Billen, a specialist in nitrogen and its diffusion in time and space. We will come back to the importance of this work for territorial ecology later.

Other analyses focused on the polluting aspect related to the important extraction of material from the subsoil or biomass. If the extracted materials are not in themselves polluting, this work focuses on materials extracted jointly, in quantities such that they can damage ecosystems[2]. Let us not forget the important use of lifecycle analysis in order to measure the systemic impact of production, the particularity of which is that it does not focus on measuring flows from a geographical point of view, but from the point of view of the production process.

The second strategy, known as dematerialization, involves developing methods for quantifying all material flows circulating within a given system. The analysis of this metabolism, for example, as conducted in the case of the Republic and Canton of Geneva (2005), enables us to identify the most critical resources, in terms of dependence on the outside world. Geneva, supported in this by the ICAST consultancy firm founded by Suren Erkman, focused on a few flows considered to be priorities (energy, building materials, food products, metals, plastics, wood and water). In doing so, it was possible to identify the criticality with respect to these resources, especially building materials, and then to propose ways to reduce this criticality (such as an effort to produce concrete from rubble to be recycled).

Other methods aim to be more exhaustive, with the objective of giving a general view of the materiality of a city, country or region, based on the aforementioned Eurostat guide. These methods are prized by authors inspired by work in urban ecology, such as Sabine Barles, and it is therefore not surprising that this is a strong methodological benchmark for territorial ecology.

---

2 The same applies to red mud extracted with bauxite to produce aluminum (Ayres *et al.* 2001).

2.1.2. *The implementation of the Eurostat method at the territorial level*

Since 2001, the Eurostat method (published in a guide by the European statistical service) has been heavily mobilized. The method was initially designed on a national scale and therefore applicable to various European countries. Figure 2.1 shows us that the initial concept is extremely simple.

A country is analyzed through the processes of physical transformation of what society achieves from the resources taken from the environment, particularly through its economic activity. Since society is not inward-looking, it also imports and exports material in more or less balanced exchanges with other countries. The analysis also assumes that a country is charged not only the quantities expressed in tons of the products it imports, but also the equivalent weight of all the materials consumed in its production (these are the indirect flows). This last point is important because it would otherwise be too easy for a country to show a "virtuous" balance sheet, in terms of environmental impact, by simply importing the most polluting goods and services to produce, which it nevertheless consumes.

This analysis is generally carried out over a period of 1 year, which implies that the contribution to the material accumulation corresponds to materials and consumer goods with a life expectancy of more than 1 year. This corresponds to materials found in household appliances, computers or vehicles. However, these quantities are negligible compared to the materials incorporated in buildings and infrastructures (bridges, roads, networks, etc.) which correspond to the vast majority of the material accumulated within a country.

While the exercise itself is interesting, given the environmental issues and the approaches taken both in industrial ecology and subsequently in territorial ecology, the scale of the entire country appeared to be ineffective. The method was very quickly applied on a smaller scale, in France on a regional and departmental scale, administrative scales on which researchers are likely to find data.

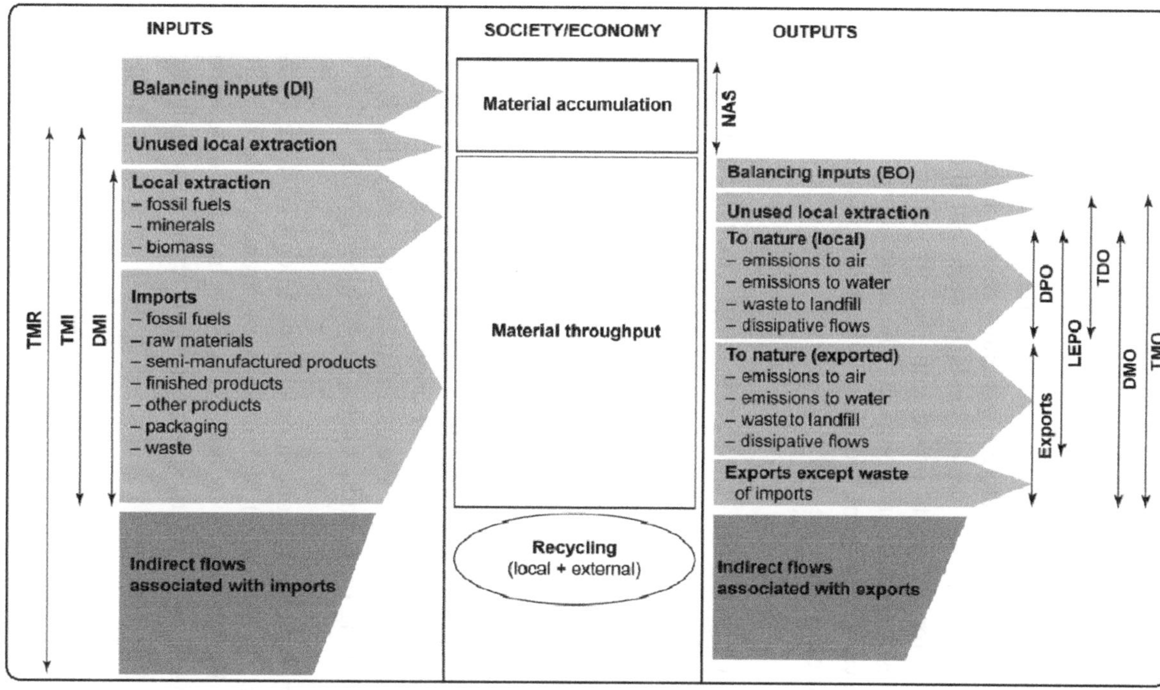

Figure 2.1. *The Eurostat method of material flow analysis (source: Barles 2009)*

In France, it is Sabine Barles who initiated this work with a study on Paris and its region (Barles 2009). In this work, Barles succeeds in adapting the method at an infra level, showing that the data necessary for the deployment of the Eurostat method are, for the most part, accessible. As the author points out, the main consequence of this downscaling lies in the nature of the flows that circulate within the system, or between it and the outside world. Thus, while country-wide flows of construction materials generally remain internal flows, this is much rarer for a city, which tends to seek out the sand or gravel necessary for cement production in its relatively close, but nevertheless external, environment. The same is true for certain waste or wastewater treatment facilities, which are frequently relocated a few kilometers from the city.

2.1.3. *A tool for analysis and comparison between territories*

In addition, this work is interesting in that it allows us to compare, for the year 2003, the metabolism of Paris and its inner suburbs, an extremely dense zone, with the Ile-de-France region, which is certainly dense as a whole, but much less so as one moves away from the capital. We can therefore see that a densely populated area logically extracts much less matter from its internal environment and imports much more (which corroborates the fact that a large city has less material autonomy). We can also see that a city like Paris exports many polluting flows and recycles very little material.

Other work carried out in other departments (Ariège, Aube, Haute Garonne, in particular) and regions of France (Midi-Pyrénées) has enabled Barles (2014) to understand the benefits of comparing the metabolisms of very varied territories, in terms of population density and land use.

This type of representation is interesting for several reasons in order to compare territories (which are otherwise similar in terms of the very low weight of recycling). We have confirmation that the addition to stock (essentially building materials) is all the more important as the territory is less densely populated and urbanized (in

Ile-de-France, 96% of the inhabitants live in urban areas compared with 67% in Midi-Pyrénées, the urbanized surface area represents 54% of Ile-de-France compared with 61% of Midi-Pyrénées). This contribution to stock is globally equivalent to urbanization and soil artificialization, one of the aspects long neglected in terms of resource scarcity, but which has been the subject of public policy for several years.

The comparison between these two regions does not stop there and concerns public policy issues. By focusing more specifically on building materials, Barles (2014) shows us that the issues are not the same, depending on the metabolic characteristics of the territories under consideration. Thus, in the Ile-de-France region, the volume of construction materials imported in 2003 (18.5 million tons) is of an order of magnitude comparable to the volumes going to landfill (17.2 million tons), bearing in mind that local extraction (17 million tons) is partly exported (6.8 million tons). In other words, an ambitious policy for recycling construction waste would likely very substantially increase the autonomy of the Ile-de-France region in relation to its needs. In the Midi-Pyrénées region in 2006, the question is quite different since, faced with local extraction of 31 million tons (and 3.7 million tons imported for 4.1 exported), "only" 3 million tons go to landfill. In other words, in this second case, a recycling policy cannot be such a major issue, when the bulk of local extraction is added to the stock (27.6 million tons). The focus of public policy in this case should deal more with a possible reduction in soil artificialization.

Another important outcome is what happens to these building materials in these various contexts. Thanks to work in various departments, it appears that if, in Paris, we consume two tons to produce one m^2 of construction, we consume much more in rural areas (6.6 tons per m^2 in Ariège and 8 tons per m^2 in Aube!). The differential is obviously not related to the construction of the buildings themselves given the standardization of construction methods. It is attributable to the infrastructure that accompanies the production of buildings to connect it to the rest of the territory. The more one builds in a geographically isolated area, the more roads, pipes and other networks are needed, i.e. the more materials are consumed. In other words, this is a measure of urban sprawl.

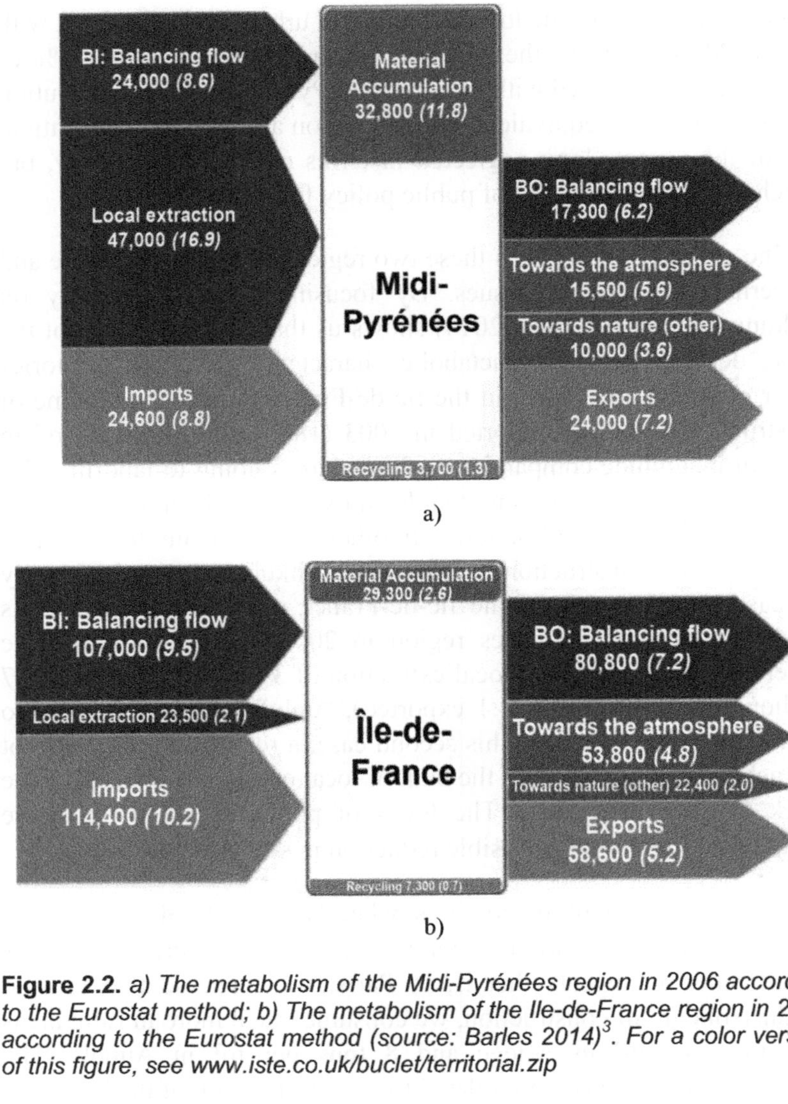

Figure 2.2. *a) The metabolism of the Midi-Pyrénées region in 2006 according to the Eurostat method; b) The metabolism of the Ile-de-France region in 2003 according to the Eurostat method (source: Barles 2014)*[3]*. For a color version of this figure, see www.iste.co.uk/buclet/territorial.zip*

3 Figures correspond to volumes expressed in thousands of tons, with the figures in brackets to those same volumes in tons per capita. The balancing flows at the input correspond globally to oxygen consumption, while at the output they correspond globally to CO_2 and H_2O emissions. Local extraction corresponds either to locally produced biomass or to extraction from the Earth's crust. In gray, we have the releases to nature in liquid or solid form, and to the atmosphere in gaseous form.

We can therefore see that implementing MFA on a smaller scale is likely to shed light on the issues specific to each territory. This work contributes to contextualizing the challenges of mobilizing the material, through "photographs" of the functioning of a territory's metabolism, and would likely help decision-makers, in terms of reflection, in order to better differentiate public policies according to the context.

On the other hand, experience has shown that the presentation of such metabolisms to local stakeholders, while arousing their interest, is far from sufficient to initiate real collective territorial project work. It is therefore certainly a relevant diagnostic tool but it remains silent on certain issues, particularly on the role of stakeholders in this metabolism.

2.2. The rise of social metabolism and the territorial footprint

As indicated in Chapter 1, social ecology emerges from the method of metabolism, inspired by industrial ecology. Social metabolism differs, above all, from industrial metabolism in that it would be used to analyze human societies and not specifically the industrialized parts of them. The work on the metabolism of human societies within the framework of social ecology is manifold, proof of a strong dynamic of questioning and of a very rich method in terms of its capacity to help us understand the interrelationships between society and nature. On the other hand, in France, a similar and very strong movement, notably under the impetus of Gilles Billen and Sabine Barles, is contributing to the multiplication of works linking issues of flows and underlying issues across large cities or territories.

A common feature of this work is the consideration of both the temporal and spatial dimensions, in order to identify the historical processes leading human societies to function according to a certain type of metabolism. Indeed, over time, human societies construct a certain type of relationship with their environment, whose metabolism is reflected in terms of material flows. In other words, the methods developed intersect with advances in history of the environment, to better understand these relationships that allow us to speak of co-evolution between nature and societies.

2.2.1. *The different facets of the place of energy in metabolism*

In her doctoral thesis, Eunhye Kim examines the evolution of the energy metabolism of the city of Paris in the 19th and 20th centuries. More precisely, the research involves evaluating all the energy consumed to supply Paris for all uses. It is therefore an exhaustive assessment that is established, in particular, to understand not only how much the city consumes, but also where the energy supplies come from, in what form (wood, agricultural biomass, oil, etc) and in what quantity.

This represents an impressive effort, aiming to identify all existing data between 1801 and 2006 in order to account for the entry of fuel into Paris, but also secondary energy (i.e. already transformed outside the geographical perimeter concerned). The result is an ability to characterize the evolution of resource consumption, which shows us a spectacular surge following the Second World War.

Beyond this significant increase in energy consumption, this work also makes it possible to establish a link between Paris and the energy-supplying territories, a link that is changing with the nature of the energy sources mobilized. There is a shift toward ever more distant energy sources, which is both a consequence and a cause of the implementation of ever more efficient transport infrastructures and networks for the circulation of material and energy mobilized on an ever larger scale.

Similar work has been carried out in Austria by researchers at the Institute of Social Ecology but on a national scale. Eunhye Kim's work thus represents a breakthrough in the methodological capacity to deal with the circulation of flows, in this case energy providers, on a territorial scale.

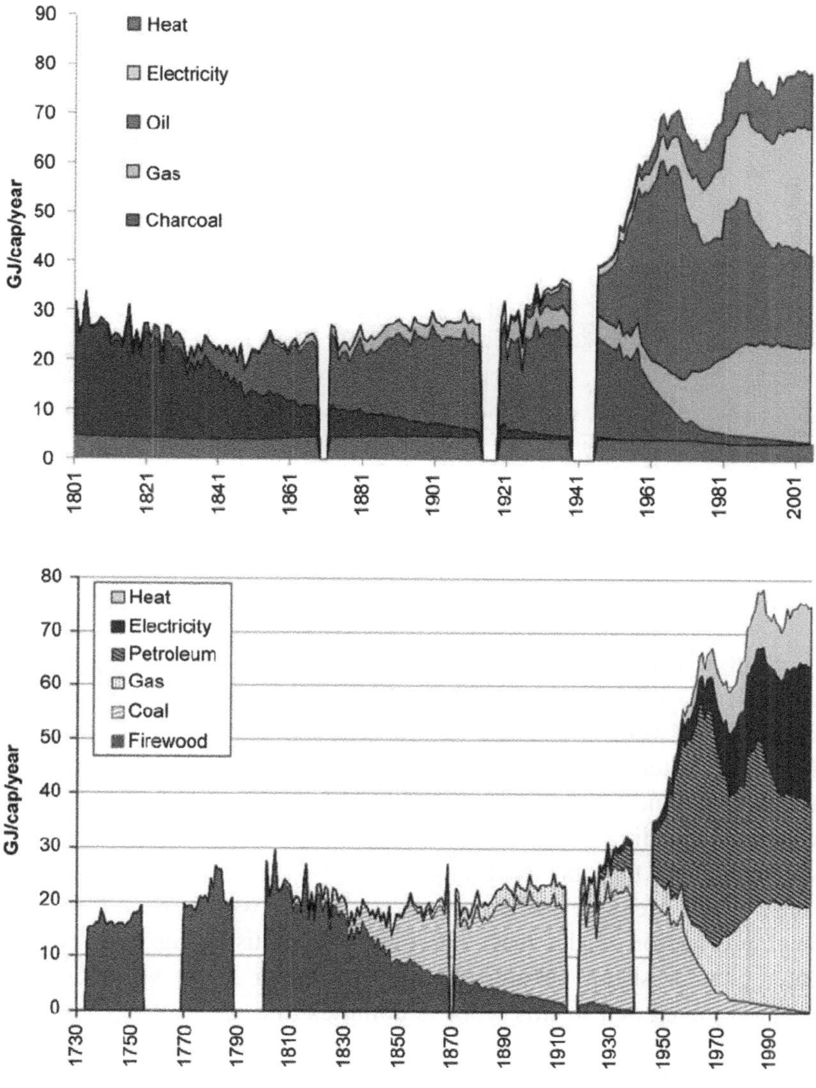

Figure 2.3a. *Final energy consumption of Paris in GJ PCS/capita/year between 1801 and 2006 (source: Kim 2013). For a color version of this figure, see www.iste.co.uk/buclet/territorial.zip*

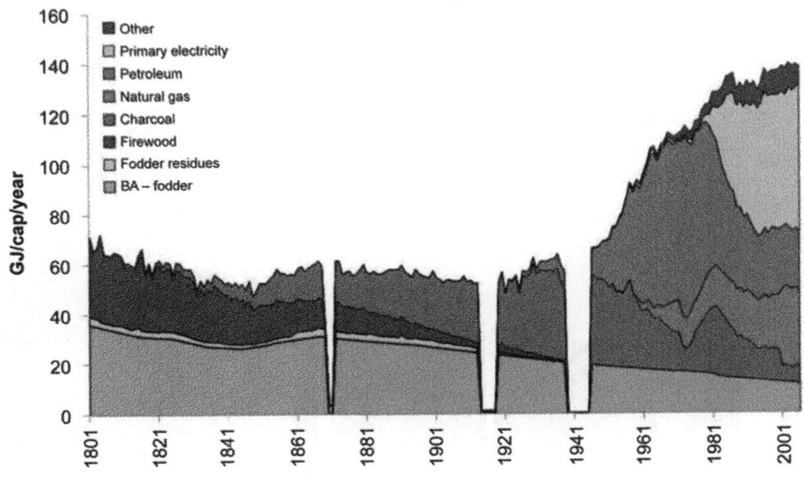

Figure 2.3b. *Overall energy demand of Paris in GJ PCS/capita/year between 1801 and 2006 (source: Kim 2013). For a color version of this figure, see www.iste.co.uk/buclet/territorial.zip*

Souhil Harchaoui and Petros Chatzimpiros conducted equally original work on the energy metabolism of a territory. These two researchers were interested in the energy efficiency of livestock breeding, in terms of the conversion of livestock feed into human food (Harchaoui and Chatzimpiros 2017). This work makes it possible to consider feeding issues in terms of energy efficiency and, consequently, the surface areas required to feed livestock, whether these areas are in France or the rest of the world. Unsurprisingly, beef production is the least efficient way of using agricultural land after goat or sheep farming (farming at much lower volumes). The method developed in this work is complex and very interesting, even if it seems to stray slightly from one of the concerns of territorial ecology, with results obtained at the country scale, taking relatively little account of the local production context. Certain questions nevertheless deserve to be contextualized. For example, can we consider the rearing of cows on fertile land (and therefore mobilizable for much more energy-efficient cereal production) in the same way as the rearing of sheep on land that is difficult to mobilize for other crops? In spite of this limitation, this work is important in that it allows new methodological advances to be established, which in the coming years may be applied at the territorial level.

2.2.2. *The central place of the foodprint*

In our opinion, the special issue of the journal, *Regional Environmental Change* published in June 2012, following a workshop in Paris in 2009, constitutes a key moment in the affirmation of work on historical analysis of the territorial footprint of cities. The analysis of the trajectories of a certain number of large and very large Western cities (Paris, New York, Barcelona, Vienna, Brussels, Athens, etc.) shows that their development is not very homogeneous, which can be explained, in particular, by the relationship of these cities to their immediate environment and the geographical conditions to which they are subject[4]. Food is often at the heart of the work (as well as access to water, as we will see later), which leads to development of the foodprint of large cities such as Paris or New York, with a variety of complementary approaches that make it possible to spatialize this foodprint over time.

Thus, Billen *et al.* (2012) are interested in feeding through the most constraining flows in agriculture, particularly in temperate zones; a flow that is essential for protein intake, i.e. nitrogen. By analyzing this flow, the authors manage to trace a large part of the foodprint of Paris, as well as its relationship to the Parisian basin, without which the city could not have expanded. Based on the diet of Parisians, as well as the average amounts of nitrogen contained in the various types of food consumed, it was possible for these authors to establish the overall dietary requirement expressed in nitrogen.

One of the main points of this work is that it makes the link between agrarian practices and the development of the city. This makes it possible to highlight: (1) from the point of view of rural areas, the productive surplus capacity of farms near Paris and, therefore, the potential of urban areas that it is possible to feed; (2) conversely, from the urban side, the surface area needed to feed

4 But also with the historical reasons for the foundation of these cities.

the city according to its demographic evolution; and (3) it follows that it is possible to analyze the degree of dependence between rural and urban areas and to establish the foodprint of Paris over the centuries.

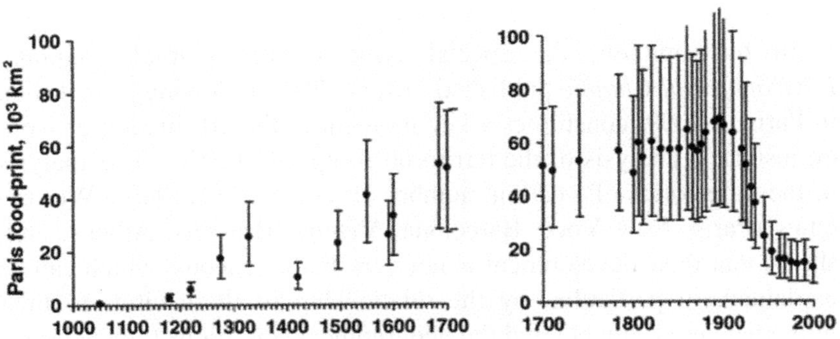

Figure 2.4. *The foodprint of Paris expressed in thousands of km² (source: Billen* et al. *2009, p. 20)*

This representation allows us to note an important increase, though not linear, in the need for agricultural land between the 11th century and the 19th century, then a marked reduction in the area needed since the last century. We can see that the foodprint of Paris depends mainly on its demographic dynamics and technological progress in agriculture, with both effects more or less compensating each other over time[5].

The work of Gilles Billen's team goes even further since, with the help of an important work combining the analysis of trade statistics in historical archives and the establishment of rules for converting tons of food into tons of nitrogen, it was possible to spatialize the foodprint of Paris (Billen *et al.* 2012b) over several periods.

In Figure 2.5, not surprisingly the foodprint of Paris at the end of the 18th century, in terms of bread and cereals, is extremely local. The average distance is 110 km, even lower for fruit and vegetables

5 Error bars are sometimes large, but they do not hinder the visualization of trends.

and barely higher for animal products. We might think this is the classic pattern, that cities naturally tend to feed themselves from their hinterland. While this statement is not entirely false, it nevertheless requires some serious revision. Thus, Derek Keene shows us that London has very quickly followed a different food trajectory. If, at the beginning of the 16th century, the grain supply area of Paris did not exceed 65 km, it was already more than twice as large for London, even though it was half as populated at the time (Keane 2012, p. 268). Even at the end of the 17th century, when Paris was four to five times the size of London in 1300, cereals came, only exceptionally, from supply basins more than 60 km from the French capital.

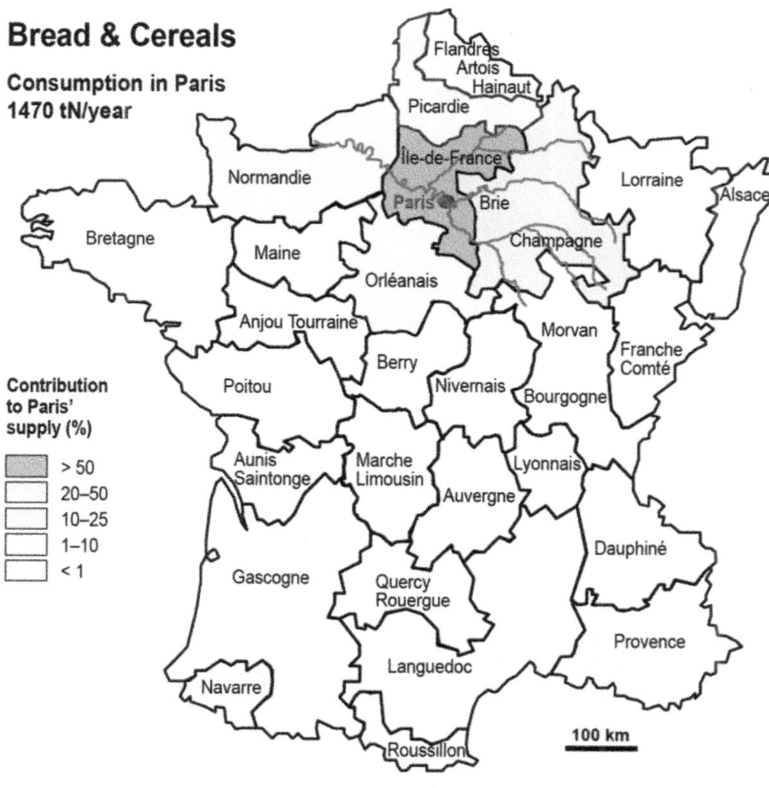

Figure 2.5. *Contribution of the French provinces to the supply of bread and cereals to Paris in 1785 (source: Billen* et al. *2012b, p. 328). For a color version of this figure, see www.iste.co.uk/buclet/territorial.zip*

The comparison is interesting and shows us that the footprint of a city does not only depend on its size, but also and largely on its geographical configuration, as well as the commercial links woven throughout history. Thus, Paris has some of the most fertile land in Europe (Beauce and Brie) close by, while London has transport facilities, both river and especially sea. Moreover, the work on Paris confirms a relative localism in terms of supply, even at the beginning of the 21st century (see Figure 2.6). Cereals travel, on average, about 500 km when London obtains its supplies from all over the world, including Australia and other distant Commonwealth countries.

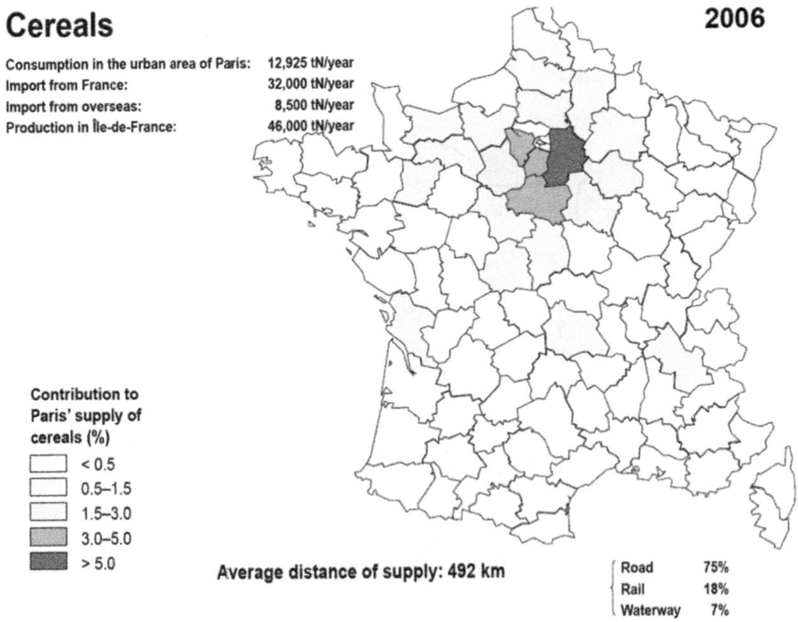

Figure 2.6. *Contribution of French departments to the supply of cereals to Paris in 2006 (source: Billen* et al. *2012b, p. 330), For a color version of this figure, see www.iste.co.uk/buclet/territorial.zip*

Foodprints are therefore more or less local, and depend on the historical trajectory of cities, conditioned by the geographical and political context. Thus, New York City was born for commercial reasons (it was originally a Dutch trading post) and not because of fertile land. By developing a method to highlight New York's "foodshed" (analogous to the work on the city's watershed), and therefore with methods that are somewhat different from those of the foodprint (although still based on measuring nitrogen consumption), Swaney *et al.* (2012) make it possible to map the supply locations of the major metropolis and see how they have changed over time. The work shows us the crucial role of infrastructure in the development of New York. For example, the opening of the Erie Canal in 1820 allowed the city to obtain supplies from the Midwestern United States, and thus to feed a much larger population.

This work is important both for the methods used, which combine methodologies specific to biogeochemistry, history and geography, and for the ability to concretely highlight the fairly strong circularity of cities. If it is understood that, beyond a certain size (which seems to depend on context), cities cannot envisage feeding themselves, it is interesting to better identify the degree of potential autonomy according to the geographical, pedo-climatic and historical context. This makes it possible, in the face of current sustainability challenges, to better situate what can be expected in terms of territorial strategies.

2.2.3. *... as well as the place of water*

Water is the other vital flow, often analyzed from the city's footprint. The urban world has mostly developed according to the availability of water. While today costly seawater desalination techniques open new horizons in terms of urban planning (but not necessarily in terms of sustainability), in the past, water has been one of the major constraints to urban population growth. Determining a city's water footprint is therefore as important as its food or energy footprint. The methods deployed to determine this are quite similar to those developed in the area of food supply. They are even simpler in terms of the quantitative dimension, as the flow concerned is homogeneous. On the other hand, the issue of water cannot be fully grasped without taking into account the quality of the flows arriving into or leaving the city.

Very interesting historical approaches have been adopted, highlighting the spectacular increase in the water footprint of large cities. This is the case of Athens, a Greek city whose water autonomy within its walls is just a memory from as early as the 6th century BCE and which, thanks to its military-to-commercial enrichment, started building aqueducts early enough to enable it to cope with its demographic growth (Stergiouli and Hadjibiros 2012). The method used mixes hydrological, historical and geographical considerations around the central notion of watershed. In particular, it makes it possible to map the area of the country needed to supply water to Athens, where about a third of the Greek population lived before the 2008 crisis and the terrible bloodshed to which the country was subjected. The work of Maria Stergiouli and Kimon Hadjibiros perfectly illustrates the spectacular increase in the capital's water footprint on the country during the 20th century.

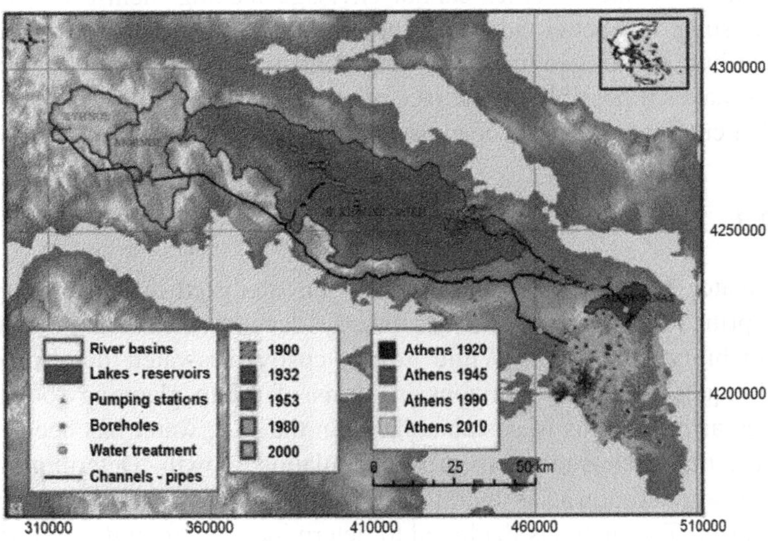

Figure 2.7. *Evolution of the water footprint of the city of Athens during the 20th century (source: Fivos-Sargenti* et al. *2019). For a color version of this figure, see www.iste.co.uk/buclet/territorial.zip*

From 1922 (reception of Greek refugees from Asia Minor) the population of Athens increased significantly (from 500,000 to 3.5 million inhabitants in 80 years); the demand for water increased even more (from 4.8 hm^3 in 1922 to about 100 hm^3 in 2000). Notably, the water consumed travels up to more than 250 km, close to the Adriatic Sea, proof of an aquatic footprint that some consider to be excessive. Others, however, believe that such a large footprint is justified in that a good third of the Greek population lives in Athens.

Poor water treatment infrastructure is the flip side of Athens' water footprint. The treatment system, with a single wastewater treatment plant (in operation since 2012), is also in the European Union's sights. Greece is regularly condemned by the European Court of Justice for failing to comply with wastewater regulations. In this sense, the water footprint of the Greek capital is oversized, given the polluting impact caused.

The parallel between Athens and New York is interesting in this sense. The same team that characterized the foodprint of New York has done important work to determine its water footprint. The watershed concept is also at the heart of the methodological framework, and makes it easy to establish the water supply regions thanks to the numerous historical sources (Swaney *et al.* 2012). Since the establishment of a trading post by the Dutch East India Company and a population of a few thousand inhabitants in New Amsterdam, the water needs of the largest American metropolis have increased significantly. Historical work shows that, contrary to the case of Athens, the issue of water supply is also qualitative (the cholera epidemics of the 19th century in New York) and concerns other issues (such as fires, which ravaged the city in 1776). Water, although very present in this port metropolis, is considered scarce and of poor quality. It has to be fetched from more distant watersheds (up to about 300 km), although proportionally quite close given the size of the metropolis and the country. The New York trajectory differs from that of Athens in several respects. In the 1930s, the necessary expansion of its water supply caused a legal conflict with other large cities and between the states of New York, Pennsylvania and New Jersey. A court decision granted rights to New York City but showed that the expansion of the city's water footprint was in conflict with that of other large cities. In fact, awareness of the problem (water consumption per inhabitant rose from 200 m^3 per inhabitant per year to more than 250 m^3 between 1930 and 1970, for a population that grew from just under 5 million to nearly 8 million) eventually led to a decline. The introduction of individual water meters and awareness-raising campaigns from the 1980s onward encouraged a decoupling of demographic trends from water consumption. In addition, major investments have led to a reduction in line losses during the transport of water from the Cannonsville reservoir. The peak of 290 m^3 of water per inhabitant and per year reached in 1988 was followed by a significant reduction in consumption, down to about 180 m^3 in 2010 (Swaney *et al.* 2012, p. 377) and even 172 m^3 per year in 2018[6].

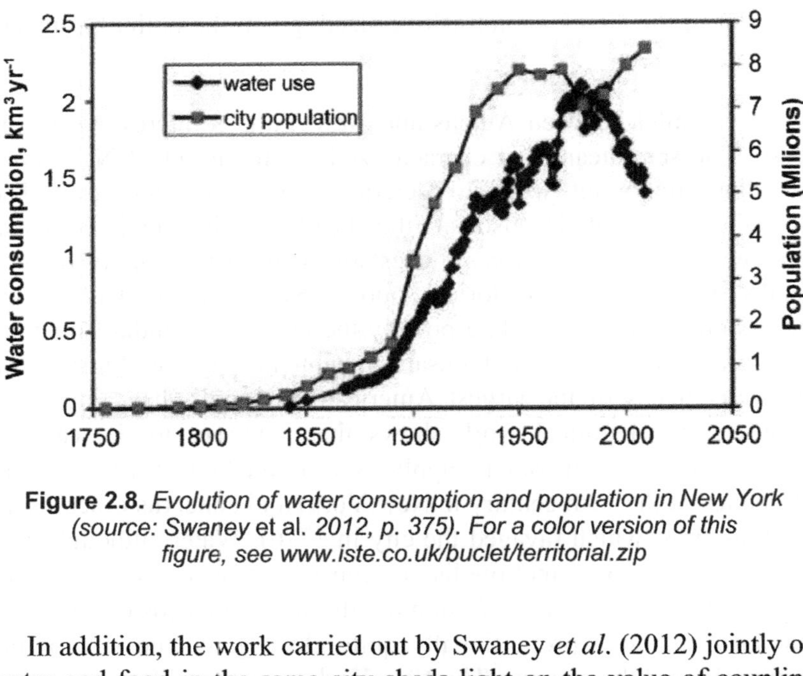

Figure 2.8. *Evolution of water consumption and population in New York (source: Swaney* et al. *2012, p. 375). For a color version of this figure, see www.iste.co.uk/buclet/territorial.zip*

In addition, the work carried out by Swaney *et al.* (2012) jointly on water and food in the same city sheds light on the value of coupling the two issues and on possible synergies. The nitrogen-based approach to the food supply, an approach shared by other researchers, as already shown, makes it possible to make the link between the pollution of rivers and water tables and, consequently, of seas and oceans (even if

6 Available at: https://data.cityofnewyork.us/Environment/Water-Consumption-In-The-New-York-City/ia2d-e54m (accessed 13 February, 2020).

the dilution of nitrogen flows in such bodies of water has, for a long time, made the issue invisible). Production based on massive nitrogen input, as practiced in industrialized agriculture, results in pollution of these environments. Even though New York eats only very marginally locally (i.e. within a radius of 200 km), local agriculture has long posed problems in terms of nitrogen loss from the soil and, as a result, the quality of the city's water supply has been called into question. Faced with water that was becoming increasingly unfit for consumption, New York City officials took up the issue and had to consider building a water treatment plant upstream. Faced with the scale of the investment (approximately 4 billion dollars) and future operating costs, a reflection was carried out, to which the work we present here, as well as other academic works, have contributed by making the link, via nitrogen, between food and water. After years of tension between New York City and the farmers in the Catskills, where much of the water consumed comes from, the city and the farmers reached an innovative agreement called "comprehensive agricultural planning"[7]. Thanks to this agreement, which began in 1990, agricultural practices were adapted to substantially reduce polluting discharges, with New York City paying for the investments in the plan. In doing so, the city is paying for the ecological services associated with agriculture compatible with the preservation of quality water, and estimates that it has saved billions of dollars compared to an end-of-pipe solution. Finally, in a context where a growing number of city dwellers are questioning the origin and quality of their food, this program promotes closer ties with New York State agriculture involved in this agreement.

We have placed emphasis on this example to show how much the concerns of work on the territorial footprint of cities echo growing societal concerns. A better understanding of the circulation of flows between a consumer city and areas that supply energy, water or food is important both to understand the issues involved and to bring these issues to the attention of public decision-makers, and sometimes even to identify certain courses of action to consider.

---

7 Available at: http://eau-iledefrance.fr/doc/wp-content/uploads/2014/04/dossierNYwater.pdf (accessed 13 February, 2020).

2.2.4. *The stakes of social metabolism*

Social metabolism represents the body of work carried out by researchers, with explicit reference to *social ecology*. The dual term is however older, since it is attributed to Karl Marx in Capital (Fischer-Kowalski and Weisz 2016), but its resurgence dates from the 1980s, as we saw in Chapter 1. Social ecology is very strongly based on a socio-ecological, ontologically naturalistic approach (Descola 2005), according to which society intervenes in nature (through labor, technology and capital) to modify it according to its needs (Fischer-Kowalski and Weisz 2016, p. 22). The authors insist on the importance given essentially to material, energy and land-use issues.

Viennese researchers also refer to the notion of socio-metabolic regimes (Krausmann *et al.* 2016) as synonymous with social metabolism but with a different intention. By focusing on the notion of regime, it highlights the temporal dynamics of social metabolism, in other words the historical analysis of the metabolism of socio-ecosystems. As we have seen from the work on the food or water footprint of large cities, this temporal dimension cannot be avoided when the issue at stake is that of reflection on questions of sustainability or transitions. The ambition of work on socio-metabolic regimes is stronger, as it involves characterizing the main stages of human history through material and energy analyses of societies (Krausmann *et al.* 2016, p. 64). Variables such as demographic evolution are gaining in importance (as we have also seen in Gilles Billen's work), but what characterizes a regime above all here is the way it mobilizes the use of materials and energy in an interrelated way. The sources of energy mobilized and especially the technologies allowing this mobilization strongly condition the possibility of moving and transforming material. Polluting emissions to the environment will also be very different depending on the energy regime.

Based on these presuppositions, the authors paint a picture of the main regimes found in history. In particular, they consider that the transitions from one regime to another are the result of revolutions (the most important of which are indicated as the Neolithic revolution, transforming hunter-gatherers into farmers, and the Industrial

Revolution, shifting from agrarian industrial regimes). The work carried out therefore aims to establish the metabolic profile of these three main regimes in history, particularly from the point of view of environmental impacts and sustainability.

In terms of sustainability issues, questioning the intensity of land use is often at the heart of the approach. It is when Krausmann *et al.* (2016) describe the agrarian regime; it is also at the heart of other work on social metabolism (see, for example, Erb *et al.* 2013). The modalities of this use would determine the sustainability of human societies, since, in the name of sustainable development, biomass products are multiplied, and this intensification is not without important ecological consequences. It is in this sense that Erb *et al.* (2016) consider it relevant to use the notion of ecosystem services as an interface between the ecological preservation of environments and the utilitarian potential of land use. To do so, the authors rely on the concept of Human Appropriation of Net Primary Production (HANPP) proposed by Vitousek *et al.* (1986) in order to measure the impact of human activity on biosphere processes. Net Primary Production (NPP) is the energy available to other living organisms once the plant has consumed the part necessary for its survival. In more recent work, biomass harvesting and productivity losses (or gains), related to land use and land cover changes, are also included in HANPP (Erb *et al.* 2016, p. 110). In particular, this calculation method makes it possible to monitor the effects of land use changes from both a socio-economic and an ecological perspective. From a socio-economic perspective, HANPP can express the need for NPP in the production and consumption of goods by human societies. From an ecological perspective, HANPP highlights the difference between the potential energy flow from ecosystems and the energy flow remaining in the ecosystem after land use change and post harvest. In other words, it is the energy that is not monopolized by humans and therefore available to other living species (Erb *et al.* 2016, p. 111). It is therefore not only a question, as in other approaches, of simply calculating the potential for exploitation of ecosystems by human societies, but also of taking into account the needs of other living species, failing which the ecosystem human activities are based upon is impoverished.

For researchers in *social ecology*, social metabolism is therefore analyzed along two main lines, sometimes worked separately and sometimes jointly: the evolution of socio-ecosystems over time and the interweaving of social and ecological phenomena. This, in particular, justifies the recognition of contextual differences from one society to another, and thus the making of comparisons between case studies.

In order to carry out this major project, a large number of methods, from many disciplines, are being mobilized or are considered as needing to be mobilized; in particular all the work dealing with the dynamics of the systems of actors. Progress in reflection on the concepts and theories to be mobilized is important. However, if there is one criticism to be made of this impressive and extensive research program carried out by the University of Vienna, it is of the scale of analysis, not from a spatial point of view, but from the point of view of humans. In the following sections, we emphasize, in particular, the few complements that territorial ecology can bring to *social ecology*.

2.3. Contributing to better knowledge of territorial issues

We have seen in the preceding sections the wide variety of methods deployed in a teeming interdisciplinary field. This variety reflects quite diverse objectives. Some researchers produce quality research without trying to identify themselves with one school of thought. Other researchers integrate themselves into schools of thought and broad interdisciplinary research programs. This is particularly the case, as we have seen, for the *social ecology of* Vienna. However, despite the impressive diversity of work carried out within the framework of the latter, it seems to us that a territorial ecology, as we understand it, makes sense and can provide both new elements of knowledge and ways of grasping sustainability issues. For us, therefore, it is a question of complementarity between the approach that we intend to favor and these numerous works. This complementarity is not developed with a view to, or rather in the face of, the mirage of research's capacity to understand human–environment relations in a comprehensive manner, but in order to increase the focus, particularly by linking knowledge and necessary action, on the major challenges of sustainability.

2.3.1. *A more participatory approach*

The advent of metabolic analysis methods is rooted in the desire of a certain number of authors to highlight that an analysis of how societies function, cut off from an understanding of the materiality of what surrounds us, was unlikely to take into account the constraints on human societies. What was true in the 19th century is all the more true as the global changes induced by our environmental inconsequence have crept into political agendas. It can therefore be said that most of the work on the metabolism of human societies is intended to remind us that we cannot understand society without taking into account the material consequences of our actions, both individual and, above all, collective. This has given rise to work carried out by quantitativists, concerned with refining methods for measuring the flows circulating within a system. Numerous results have been obtained, making it possible today to observe the metabolic functioning of a system and to measure its effects in terms of flows.

The question that then arises is that of knowledge transmission, particularly among actors concerned with modifying their behavior toward greater sustainability. We were able to observe during public meetings the undisguised interest of participants in the presentation of the quantitative metabolic functioning of "their" territory. The graphs illustrating this metabolism impressed the participants, but the appropriation of the results presented, particularly in a logic of action, was just as lacking. From this, we deduced that this work needed to be completed, or modified, in order to allow the stakeholders concerned to take ownership of the issues underlying these metabolic analyses.

To do so, a first step consists not of developing new metabolic models, but of ensuring that the representations produced really speak to the stakeholders of the territorial systems concerned. One way forward undoubtedly consists of simplifying the representations in order to focus on aspects that are accessible and that the stakeholders can understand the issues. Above all, we see two points that we would like to emphasize: the co-construction of representations and the construction of representations by issues. These two points must be combined. In Chapter 3, we will present the division of the territory into wealth-creating subsystems, as we initially defined them in a study of the commune of Aussois in Savoy (Buclet 2015b). Representing the territory on the basis of wealth creation is a way of putting the issues at the heart of the representation. It allows us not only to see what is circulating, but also what the purpose of this circulation is. In this sense, in our opinion, it has an advantage over the Eurostat method, which, though very useful, is not well suited to a dialogue with the stakeholders of a territory.

From the point of view of the participatory method, the stakeholders can initially focus on the development of the metabolism of the activity to which they are most closely linked, in order both to co-produce the representation of this activity and to provide qualitative and quantitative elements of knowledge, in order to give form and substance to the metabolism. The information can thus relate to the quantities of circulating flows, and also to the stakes involved in this circulation. Second, it is important to bring together the stakeholders in order to reflect on the relationships formed between them, based on the developed metabolisms. In doing so, the stakeholders can take the measure of the existing or missing interrelationships between the structuring activities of a territory. We refer the reader to Chapter 3 to understand what we mean by structuring activities, which we also call wealth-creating activities. However, it is important here to evoke the question of the co-construction of the metabolism of a territory, in order to make the reader aware of this necessity to think about the operationalization of metabolic work, in accordance with the stakes of the territories to which it applies.

2.3.2. *A more territorially focused approach*

What has just been said in terms of participation invites us to consider a less scientific and more transdisciplinary approach[8]. Speaking of territorial ecology rather than *social ecology* is not innocent. The anchoring of reflection to the notion of territory encourages a rapprochement between researchers and actors of the investigated territories, and the way they live their daily life and project themselves from it in the future. From this point of view, we are getting closer to the work of Alberto Magnaghi (2012) and his concept of the urban bioregion. According to this author the territory is:

> the product of processes of co-evolution between human labor and nature, these processes do not produce a total artifice, but *living neo-ecosystems*, where transformed nature assumes different balances in turn. (Magnaghi 2012, p. 10)

The transformation of nature, the evolution between nature and society, is a slow process that must be taken into account. The territory:

> is conceived as a subject, a living organism of high complexity produced by the encounter between cultural events and nature, composed of places (or regions) endowed with identity, history, character and long-lasting structure. (Magnaghi 2012, p. 10)

It is therefore important to consider the elements of the past not only in terms of polluting emissions and environmental degradation, but also in more positive terms. Human intervention gives rise to socio-ecosystems resulting from interaction with the environment, each of which is unique in itself and generates cultural specificities that are at the heart of a territory.

8 Let us recall that transdisciplinarity consists, beyond multidisciplinarity (making several scientific disciplines work together) or interdisciplinarity (crossing concepts and methods between scientific disciplines), of making researchers and non-researchers work together, or even defining research subjects together.

By providing a territorial perspective, the proposal of territorial ecology is thus particularly sensitive to the changes in the relationships between societies and nature, from the point of view of the meaning that this produces[9]. This meaning is what makes the world inhabitable and places are marked by particularities that make them endearing for the humans who inscribe their existence there[10]. Also, we can say that what differentiates territorial ecology from related interdisciplinary fields is the predominant or complementary place given to functionalism. Indeed, what characterizes metabolism is that it explains a large part of the territory, or any other analyzed system, from its functioning. Without denying functionalism, territorial ecology proposes not to give up less "mechanical" explanations of the functioning of territories. The meaning of what is done counts, in our view, as much as what is done within a territory. This posture is likely to enrich work based on a metabolic approach to human societies.

2.4. Conclusion

The concept of metabolism encompasses many ways of representing the material impact of human societies. In this chapter, a non-exhaustive overview allows us to see that the metabolisms obtained depend on several aspects:

– the perimeter taken into account, whether in terms of geographical scope or in terms of human activity;

– temporality: some works favor snapshots of the territorial metabolism, even if it means multiplying them in order to observe the dynamically analyzed system, while other works are devoted primarily to the analysis of temporal dynamics, with an emphasis on regime phases and transition phases;

9 In this sense, see the chapter in the book on Aussois by Sabine Barles, Juliette Cerceau and Marion Debuisson (2015), which insists on the transformations over time of a territory likely to explain, as *social ecology* does, changes in socio-ecological regimes. This work is based less strictly on quantitative explanations of changes in metabolism, but on a historical understanding of these changes, through demography, culture, habits, etc.

10 "We affirm that a world without places is a world uninhabitable for man" (Magnaghi 2012, p. 14).

– the meaning of the results obtained, whether it is to show what is circulating in order to highlight potential metabolic imbalances, or whether it is to insist on the reasons for this circulation.

No metabolic representation can answer all the questions that arise in understanding the relationship between humans and the environment. The complementarity of approaches is therefore relevant and it is advisable to mobilize one or the other according to objectives, whether these are purely scientific or expressed by the actors of a territory. In Chapters 3 and 4, we present a proposal for metabolism aimed, in particular, at responding to some of the shortcomings identified in this chapter. In particular, we emphasize this method by highlighting the issues of human activity, as well as by the more cultural and symbolic questions of the relationships humans weave with what surrounds them.

3

A Wealth Creation Approach for a Territorial Diagnosis Methodology

June 2013 was an important moment for territorial ecology, with a thematic school labeled CNRS on territorial ecology being run. If we use the definition given by the institution that governs these schools, the thematic school is a training modality, open to scientific communities. For the CNRS, they are one of the tools used in the institute's scientific policy and the development of interdisciplinarity. The thematic school is a means of encouraging the emergence of new themes, of accelerating the dissemination of conceptual, methodological and technical advances, of encouraging the creation of new communities, of organizing or renewing a (scientific) community and of providing assistance for thematic reconversions[1]. The aim of the thematic school held in Aussois (Savoie) was to consolidate the beginnings of an interdisciplinary scientific community around territorial ecology. During this school, the choice was made to give substance to methodological exchanges by applying these choices to an analysis of the Aussois commune. The groundwork was prepared beforehand and throughout the week; as many elements as possible to work on the metabolism of the commune were provided.

This work has made it possible both to capitalize on certain already-proven methods, such as the foodprint of a territory expressed

1 Available at: http://www.cnrs.fr/infoslabos/formation/shs.htm (accessed February 14, 2020).

in nitrogen mobilization (Bonaudo *et al.* 2015), and to propose new ways of modeling metabolism (Buclet *et al.* 2015c). It is the latter proposal, in particular, that is the subject of this chapter. Indeed, since the thematic school, held in 2013, and the publication of a collective work on Aussois in 2015, the approach to metabolism divided by wealth-creating subsystems has been expanded and aims to reinforce a number of points that have not been sufficiently addressed in other works. In particular, while several studies, as we saw in Chapter 2, make it possible to determine what circulates within a territory, or even to analyze the links between this circulation and the historical context, analysis of the dynamics of stakeholders has not been systematized. In other words, while many studies focus on the *what*, on *what circulates*, and lead to graphic representations of the material quantities exchanged within a territory or with the outside world, as well as the emissions induced by these exchanges, answers to *how* and *why* these flows are exchanged are often missing. This question is typically at the heart of the social sciences. What motivates stakeholders to exchange flows? There is, of course, the economic question, since exchanges between individuals or groups of individuals are frequently, and rightly, associated with the benefit that each individual derives from the operation, but there is much more. As Bronislaw Malinowski (1922) showed, about a century ago, with regard to shellfish exchanges within the Trobriand Islands, human societies also structure themselves with the help of exchanges that are devoid of economic motivation. To exchange is also to create links between people, to pacify relations between individuals and between groups and to favor socialization. Through the objects that circulate, there is therefore also a symbolic part that translates intentions.

All of these intentions, which may explain the circulation of materials, water and energy, deserve to be the subject of work in addition to quantitative work on flows. It is through this coupling that we can grasp the complexity of the metabolism and better explain the intentions with which, beyond work on the territorial footprint of cities (see Chapter 2), stakeholders interact with their environment. We will consider that human societies are enriched by these interactions and that they produce individual and collective well-being at the level of social groups. In order to understand how these social groups produce well-being, the link with the environment being crucial, the most relevant territorial scale is that of living areas. This scale is not without methodological problems, as we shall see, in that the statistics that can be used to quantify the circulation of flows are mobilized more on the basis of administrative breakdowns.

Starting from this idea of enrichment, the methodology outlined in Aussois, which has since supported territorial ecology, is that of analyzing the territory by dividing it into wealth-creating subsystems. Wealth is thus understood as the enrichment of human societies in interaction with their environment and not as monetary enrichment. A territory thus divided makes it possible to analyze the dynamics of the main enriching activities, in that we consider that they have a structuring role for the territory.

3.1. The model of wealth creation subsystems

The starting point is therefore the structuring activities of a territory. The determination of these structuring activities presupposes an approach based on material flows. We must question activities involving a large circulation of flows, especially material flows. A prediagnosis is therefore necessary for each territory in order to identify these activities. In the case of Aussois, we identified four: the agricultural subsystem, the tourist subsystem, the water–energy subsystem and the residential subsystem. These subsystems are quite diverse in nature. The first two are in fact fairly classic activities, those which are customary to analyze in economics and to consider as sectors. The water–energy coupling is more unusual. It corresponds to the role played, for the commune of Aussois, by the reservoirs built upstream, with a significant production of hydroelectricity and extremely positive repercussions for the commune. Finally, the residential subsystem is based on the importance, for a territory, of those residents whose income does not depend on an activity located in Aussois, but who nevertheless support the local economy, whether they are pensioners, civil servants or commuters (people who commute between their place of residence and their place of work).

3.1.1. *The starting model*

A subsystem is modeled in the following way (see Figure 3.1).

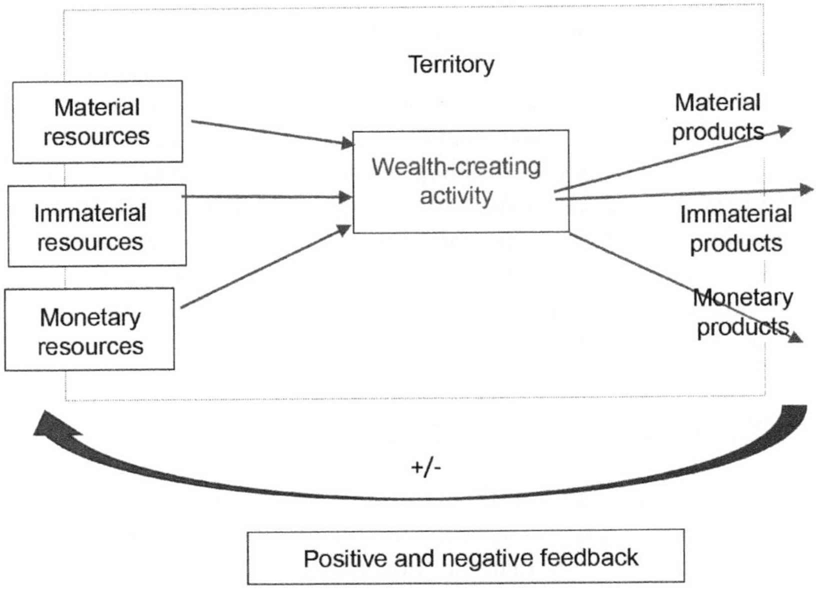

Figure 3.1. *Wealth-creating subsystem (source: Buclet* et al. *2015). For a color version of this figure, see www.iste.co.uk/buclet/territorial.zip*

In order to produce wealth, the implementation of a wealth-creating activity presupposes the mobilization of resources. Resources are stocks, the value and quantity of which change over time. For this purpose, we have, flows that are linked to the mobilization of resources entering the subsystem and flows that are produced by the implementation of operations specific to the activity leaving the subsystem. These resources may be internal and/or external to the territory, just as the products may remain within the subsystem or leave it (either to go to other subsystems in the territory or to migrate outside the territory). The products of the activity generate feedbacks on the state of the resources whose effect is positive or negative and which can increase or decrease the quality and/or quantity of the resources that can be mobilized within the activity.

There are three types of resources:

– material resources, which generate flows that are typically measured within the metabolism of a territory (substances, materials, water, energy, etc.);

– immaterial resources, which generate flows that are generally complementary to material flows (know-how, trust, etc.);

– monetary resources, such as grants or income from the previous year, that are re-injected into the subsystem to pay for certain resources (e.g. the know-how of people whose work must be paid for).

On the other hand, the activity also generates material flows (cheese, chairs, energy, pollutants, waste, etc.), immaterial flows (notoriety, reinforcement of know-how through practice, etc.) or monetary flows (income, taxes, etc.). As we can see, what is produced may be voluntary or involuntary (or at least unintended), such as, for example, carbon dioxide emissions to the atmosphere.

In the case of Aussois, and knowing that the headings of the different categories included in the model had not yet stabilized at the time of the work, this results in diagrams such as the one in Figure 3.2.

As can be seen, it was necessary to adapt the dotted line representing the perimeter of the Aussois commune to take into account the fact that many elements of the subsystem are both internal and external to the communal territory, which is not surprising, given the size of a commune like Aussois.

As in other representations of territorial metabolism, the timescale usually used is that of the year, which makes it possible to adapt the elaboration of the metabolism with the databases frequently established on this timescale. This representation nevertheless poses some problems, in the case of strong seasonality of the activities considered. This is what we were able to observe quite quickly during the exercise carried out in Aussois, in the case of tourism. Therefore, the form of representation has been adapted for the circumstances.

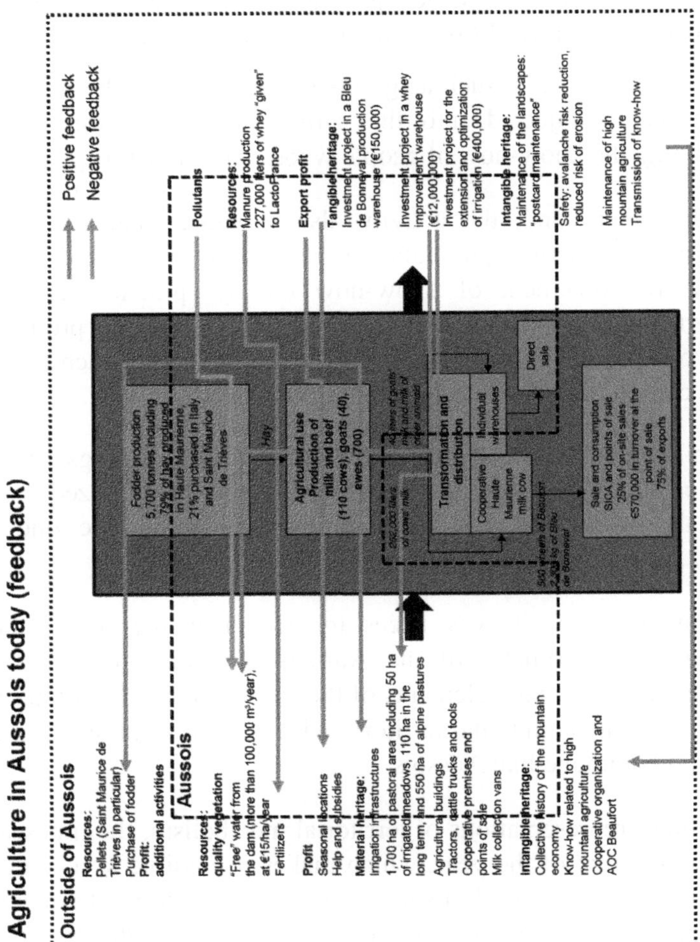

Figure 3.2. *Agriculture in Aussois in 2012 (source: Buclet et al. 2015). For a color version of this figure, see www.iste.co.uk/buclet/territorial.zip*

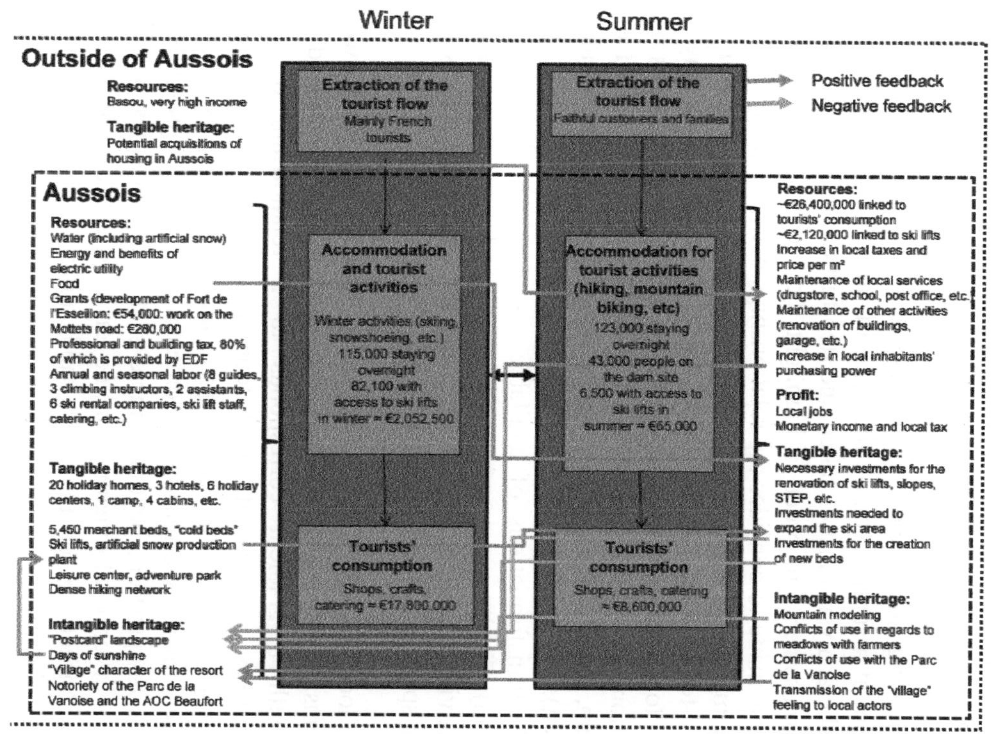

Figure 3.3. *Socio-ecological metabolism of the tourism subsystem in 2012 in Aussois. For a color version of this figure, see www.iste.co.uk/buclet/territorial.zip*

We will not go into the details of this diagram here, which we have reproduced *in extenso*. Above all, it is important to allow the possibility of distinguishing between summer and winter activities, and to represent the seasonality of activities when this affects the territory's mode of functioning.

Finally, let us recall that if the above representations are of wealth-creating subsystems, this means that the analysis must be carried out at a supra level, on the scale of the territorial system as a whole. The analysis must be taken one step further, by articulating the various subsystems in order to highlight the interrelationships. The choice was made to consider the relationships between subsystems in the same manner as relationships between living species, according to an ecosystemic logic. Along these lines, some resources may be common to several subsystems (according to a logic of mutualism or competition), or may be negatively affected by the activities of one subsystem, to the detriment of other activities (we will then speak of parasitism). Symbioses are also possible between activities, with favorable transformations of available resources (another form of mutualism). Finally, some activities may benefit from the spin-offs of other activities, without the latter feeling any particular effects (we will then speak of commensalism).

If we look at Figure 3.4, the blue refers to commensalism between subsystems, the red to parasitism, the green to mutualism and the arrow with a gradient corresponds to a relationship of both commensalism and parasitism of tourism *vis-à-vis* the residential sphere, which betrays the very frequent ambivalence of the resident populations toward the tourist phenomenon. Thus, this representation allows us to question, beyond simple sectorized activities, what is happening in terms of dynamics on the territory, and even to identify possible conflicts or, on the other hand, alliances of circumstances between stakeholders involved in such and such a wealth-creating subsystem.

This representation of the territory is well in line with the logic of the metabolism of a system, but with some notable differences. First, in many representations of metabolism (following the example of the Eurostat method), the flows that circulate are listed and accounted for, and the variation in material stocks is determined. This is primarily an accounting and material logic of flows. This accounting logic makes it possible to obtain interesting analyses, particularly in the territorial footprint analysis phase. Representation through wealth-creating activities is a complementary approach, in that it provides an economic and sociological view of metabolism, in order to clarify the way in which a territory generates wealth. We can better understand why flows circulate, and for what purposes (which, as we said above, are not only economic). In any case, this is one of the purposes of this representation, some aspects of which we will develop later.

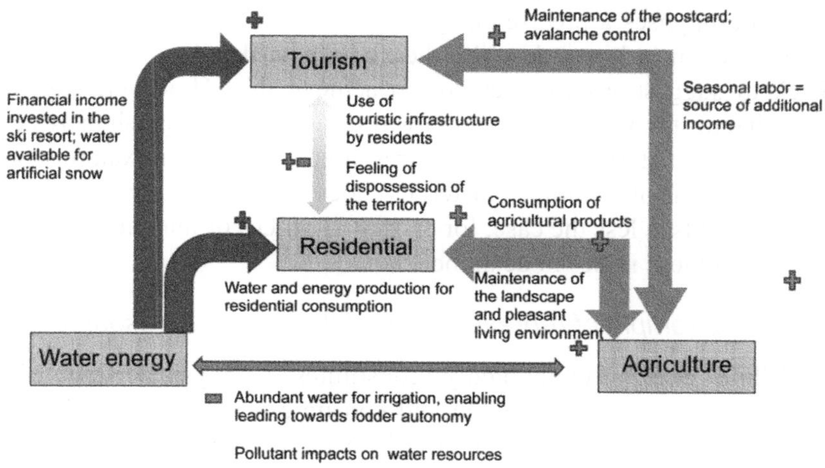

Figure 3.4. *The territory system of wealth creation in Aussois (source: Buclet* et al. *2015, p. 158). For a color version of this figure, see www.iste.co.uk/buclet/territorial.zip*

Second, beyond the why, one of the challenges of this representation is to better explain how wealth is produced by a territory. The answer to this "how" is one of the major methodological issues that enables, in our opinion, linking the concerns of social sciences with the metabolic issues of a territory, in other words, linking two complementary ways of explaining the functioning of a human society. It is this other point that will be discussed in the rest of this chapter.

3.1.2. *Multiple questions*

As this metabolic model has been grasped by researchers, doctoral students and even master's students in urban planning, a certain number of questions have emerged. The representation, as proposed, appears to be fairly simple and rigorous. However, as soon as one wishes to apply it, this simplicity is quickly called into question.

First, it is not always easy to identify all the resources mobilized by a wealth-creating activity, as well as all the products resulting from it. This presupposes a precise analysis of the activity, field surveys and interviews in order to describe the activity in detail. We will come back to suggestions for methods to achieve this. However, beyond this diagnostic work, it is not always easy to distinguish between material, immaterial and monetary resources. While, for example, the classification of building materials does not pose any particular problem, this is less the case, for energy resources or human labor. Let us look at these issues in detail below.

3.1.2.1. *Example of representation of the electrical resource*

While energy undoubtedly constitutes a physical flow, the resources required to mobilize and consume it are multiple. Let us take the case of electrical energy. In a large number of cases, this flow comes from outside the territory, since electricity production experienced significant phenomena of concentration during the 20th century, especially in search of economies of scale. It is quite rare that the key players in a wealth-creating activity produce their own electrical energy. Therefore, what resources should be taken into account at the scale of the territorial subsystem?

If the electricity comes from outside, then it is a flow whose import is made possible by the mobilization of several other resources. The first is constituted by the distribution infrastructures that make the circulation of the flow within the considered system possible, an indispensable physical network that creates a stock of material to be taken into account, with possible modifications of the stock according to the renewal of the network, its enlargement or its reduction. The second type of resource is immaterial. These are the various contracts and other agreements concluded between the stakeholders importing electricity within the considered system and the stakeholders with the power to promote this import. These are either the stakeholders who emit the flow (electricity producers) or the stakeholders who facilitate the circulation of the flow (the regulator who sets the rules for the circulation of electricity, contracts, etc.).

If we make a diagram centered on electricity, here is what it can look like (Figure 3.5).

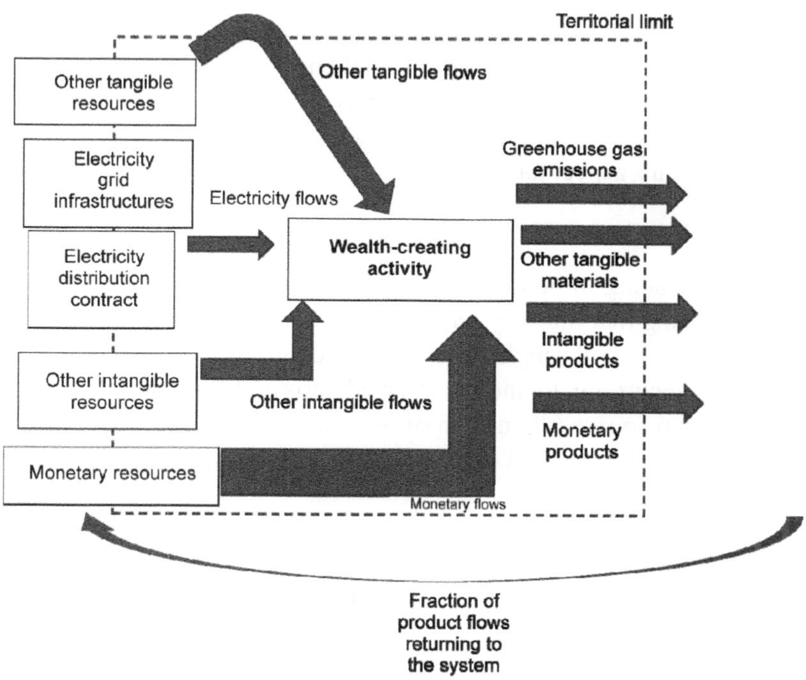

Figure 3.5. *The place of electricity in a wealth-creating subsystem. For a color version of this figure, see www.iste.co.uk/buclet/territorial.zip*

In Figure 3.5, we only show the stocks and flows, not the stakeholders. In order not to load the diagram unnecessarily, the immaterial flows linked to the facilitating stakeholders (everything that is regulatory, for example) are also absent. This type of representation makes it possible to establish the link between the material and immaterial dimensions of the same reality. Of course, separating what emerges from the use of electricity from the other operations necessary for the production of wealth through an activity makes little sense. Nevertheless, it seemed necessary to us to illustrate in more detail the way in which this is done, whether it involves electricity or other mobilized resources.

In parallel to this representation of stocks (resources) and flows (from resource mobilization) related to electricity, the question of interactions between stakeholders is also relevant, even necessary. To do this, the challenge is to reflect, for each flow, on the interactions between the three categories of stakeholders already mentioned (transmitters, receivers and facilitators) and to highlight the power relationships that can be established. We will return to this complementary aspect later.

3.1.2.2. *The place of human work*

Our bias being to be interested in wealth-creating activities, it goes without saying that the very term "activity" implies being interested in the people who, through their work, contribute to this creation. However, according to the established terminology, some questions arise. Is it a resource (the notion of human resources has been used for many years)? A stock? A flow? Is it of a material or an immaterial nature?

The consideration of humans in an essentially materialistic framework and the need to classify them in this framework raises a number of questions. Any representation of society that reduces humans to their functionalist side poses, in our view, as much of a problem as a utilitarian vision of society as imagined by economic science. Yet, on the other hand, the rest of the living world has long been treated as capital, or as a resource, without this posing any particular ethical problem.

With regard to this last argument, it seems to us more coherent, as far as human labor is concerned, to consider it on the same level as animal or plant exploitation, rather than a choice consisting of removing living things from the category of resources, which would be tantamount to denying that our human societies are part of the world of living things. Therefore, people employed within a structuring activity will be considered as a stock of material resources and their skills as a stock of immaterial resources. The activation of these skills placed at the disposal of the activity generates various flows, whether they are physical ability or information loaded with knowledge, know-how and skills that contribute to the success of the activity.

As we can see in these two examples, the great difference between what we propose and other metabolic analyses lies in this introduction of the immaterial dimensions that inevitably accompany material flows. Is this a denaturation of methods based on metabolism? The question seems legitimate to us since, by this method, we are giving new weight to the questions of a finer understanding of the relationships between stakeholders. These questions have previously tended to erase somewhat the importance of the materiality of societies and, therefore, of our relationship to the environment. This is the pitfall that we must avoid, that of favoring the analysis of the relations between stakeholders to the point of losing this materiality. In other words, the tension between materialist and relationalist explanations should not be underestimated, but, on the contrary, be regularly in the line of sight in order to arbitrate in a constructive sense.

3.2. Methods of analysis of the relationships between stakeholders within the territorial metabolism

It is therefore a major methodological challenge to complete the quantitative metabolisms of material flow balance by understanding the relationships between the stakeholders at the origin of these flows. A certain number of avenues have already been evoked in the aforementioned *Essay on Territorial Ecology*, published in 2015. We are now going to take some of these paths as a starting point in order to propose an array of tools that can be mobilized.

3.2.1. *Which stakeholders should be taken into account?*

As mentioned above, a distinction must be made between flow emitters, receivers and facilitators. However, this distinction is not sufficient in itself. A receiving stakeholder may be at the origin of the circulation of a flow (as in the case of an import) or undergo a flow (in the case of the emission of a pollutant). It therefore seems important to us to detail the categories of stakeholders and what we consider to be within them.

Therefore, in parallel with the three proposed categories, it becomes important to establish the role of each of them according to a certain number of attributes that seem relevant to us. In the work done in Aussois, we opted for a representation developed by Sabrina Brullot in her doctoral thesis (Brullot 2009), which was itself inspired by work in management sciences (Mitchell *et al.* 1997).

This representation makes it possible to articulate the various stakeholders in a project. For each stakeholder, it is necessary to determine the attributes that correspond to them among power, legitimacy and interest. The stakeholder's power is considered here to be a coercive type of power (forcing others). Legitimacy corresponds to another facet of power, based not on force but on the credit that the other stakeholders give to a legitimate stakeholder to act, propose, coordinate or even carry out collective action, or in the name of the other stakeholders. As for a stakeholder's interest in a project, this interest may be economic, political or even environmental. This attribute represents the fact that a stakeholder expects something positive from a project.

At first glance, it would seem that the projects most likely to be successful are those with a pivotal player and, therefore, a natural leader. Nevertheless, what may appear to be an asset is often a weakness. What happens when the pivotal stakeholder (whether a physical or moral personality) disappears or simply withdraws? Is there not a significant risk of breaking the dynamic? In the same way, can we not consider that there is an excess of dependence on a pivotal stakeholder and that this stakeholder takes inappropriate paths of development because they do not listen to the arguments and points of view of other stakeholders?

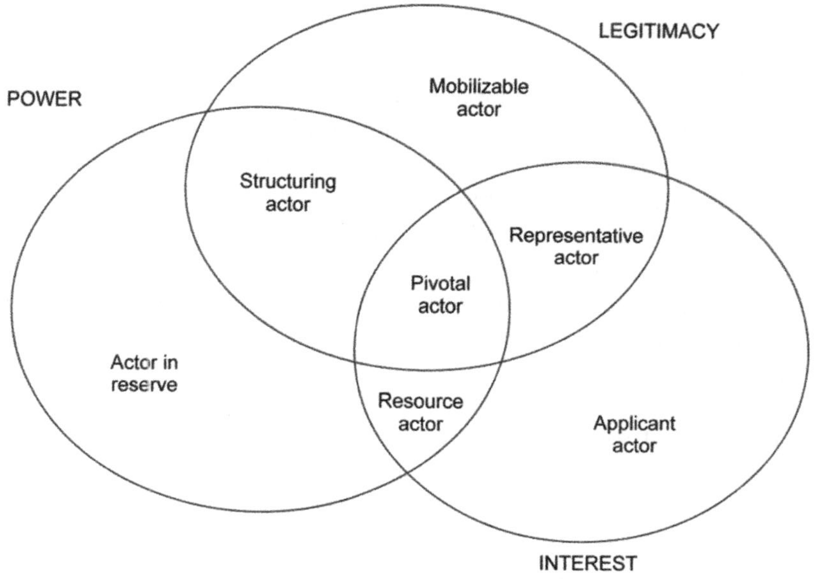

Figure 3.6. *The positioning of players within a project (source: Brullot 2009)*

Some examples of industrial ecology can shed some light on this point. Thus, in the case of Kalundborg (see Chapter 1), most descriptions of industrial symbiosis indicate that the central stakeholder is a thermal power plant for energy production. Although it is a powerful stakeholder with an interest in the process, this does not make it a pivotal stakeholder. This stakeholder does not appear to be particularly able to lead all the stakeholders in a certain direction, unlike the municipality, which undoubtedly represents the pivotal stakeholder of the territorial system under consideration. The symbiosis is remarkably stable because of the nature of the pivotal stakeholder, a public stakeholder who is *a priori* perennial and who has never failed to support and invest in local dynamics. To further increase the sustainability of the approach, a collective structure entirely dedicated to symbiosis was created about 10 years ago: The Kalundborg Centre for Industrial Symbiosis, later renamed Symbiosis Center Denmark.

Let us now examine the case of the industrial symbiosis of Grande-Synthe, within the Urban Community of Dunkirk, France. In this case, the stakeholder at the center of the symbiosis is the steel company Sollac (since acquired by ArcelorMittal), a producer of large quantities of energy recovery (i.e. produced without direct intention, and therefore generally lost and dispersed in the atmosphere). André Demarthe, the mayor of Grande-Synthe at the time the process was launched (1999), was himself a former director of the Sollac plant, which increased its legitimacy. In this sense, he could play the role of pivotal stakeholder, but not entirely, since the municipality had long been included in the Urban Community. His power was therefore restricted.

However, it soon became clear that a dynamic based solely on the will of a mayor, however legitimate it may be in the eyes of local stakeholders, could not be satisfactory, given the sustainability of the actions undertaken. This is why the Ecopal association was founded in 2001 with the explicit aim of deploying industrial ecology. This stakeholder allows the pooling of efforts between industrialists and therefore represents an interface between stakeholders with economic power and a definite interest in an industrial ecology approach, and stakeholders who have an interest in supporting the approach (the Urban Community, the port of Dunkirk, the department, the region, the Ademe, the Artois-Picardy Water Agency, the European Union). Thus, Ecopal is an interface player that brings together various players with varying attributes of power, legitimacy and interest and thus stabilizes the approach, nearly 20 years after its launch.

By studying these two examples it appears that sooner or later the creation of an interface structure can play the role of a pivotal stakeholder, a solution likely to increase the potential for the sustainability of an approach, much more than if it were based on a simple stakeholder. These interface structures allow the various players to create a dialogue, to agree on common objectives, or even on common principles that each other player must follow, at least if they intend to remain in the process.

This reading grid thus allows us to position the stakeholders according to their centrality regarding a project. As far as we are concerned here, the project is made up of the activity that creates wealth. For some, these activities have existed for a very long time (agriculture is the typical example of an activity that seems to be perennial in many regions of the world) and it may seem surprising to worry about their sustainability. There are, however, at least two good reasons to question it. The first is that with the industrialization of production processes, farms in the least fertile regions and/or with greater relief (typically mountain agriculture) have decreased and with that the agricultural surface effectively mobilized[2]. There are therefore territories where the sustainability of agricultural activity is in question.

The second reason is related to agricultural practices. Indeed, in the case of conventional agriculture, environmental degradation is significant and can lead to the abandonment of agricultural practices in the long run. Several works remind us of the importance of the phenomenon of the disappearance of arable land through erosion (see, notably, Nahon 2008) linked to bad agrarian practices.

These considerations point to the need to couple the interaction of stakeholders and metabolism analysis. Indeed, agricultural activity depends as much on the interplay of stakeholders and the economic conditions that may result in a more global context, as it does on

2 Thus, in France, between 2006 and 2014, 524,000 hectares of agricultural land were artificially farmed and 815,000 hectares became natural soil (moorland or wasteland, wooded areas). Available at: https://wikiagri.fr/articles/la-crise-agricole-en-france-expliquee-par-les-chiffres-et-les-graphiques/8601#1_reacuteduction_des_surfaces_agricoles (accessed March 19, 2020). These figures clearly show that the reduction in agricultural land is certainly due to the urbanization of the territory, but also and above all due to an agricultural decline linked to the reduced economic interest of a growing number of lands.

ecological balances that result in an increased looping of organic matter and nutrient flows. The approach proposed above has the merit not only of highlighting the importance of a pivotal stakeholder, or at least an association of several stakeholders to assume this pivotal role, but also of highlighting the issues of power relations that may exist between stakeholders. This aspect seems important to us in order to understand who decides on the circulation of flows and, therefore, who can further influence metabolic changes when these prove necessary.

3.2.2. *How to take into account the relationships between stakeholders?*

In order to make the link between metabolism and relationships between stakeholders, two complementary paths seem to be potentially fruitful. The first consists of mapping the stakeholders in relation to flows, producing a twin diagram of the stock and flow metabolism pattern indicating which stakeholders are involved in the circulation. The second is not to lose sight of an analysis of the stakeholders' motivations. For what reasons do they act? With what goals? But also in relation to what principles and values?

3.2.2.1. *Flows and stakeholders*

Figure 3.1 has allowed us to illustrate the type of figure that could be produced with the proposed approach. But who is behind the resources mobilized? And who is making the flows circulate? Who contributes to the production of wealth?

In order to provide elements of method, we decided to resume the work done a few years ago in the case of Aussois. By starting more precisely from the Beaufort production (main agricultural activity in the commune) and by focusing on the relationships of the type of main stakeholder (the cattle breeder) of the subsystem, we can draw a network of relationships between this stakeholder and other stakeholders involved.

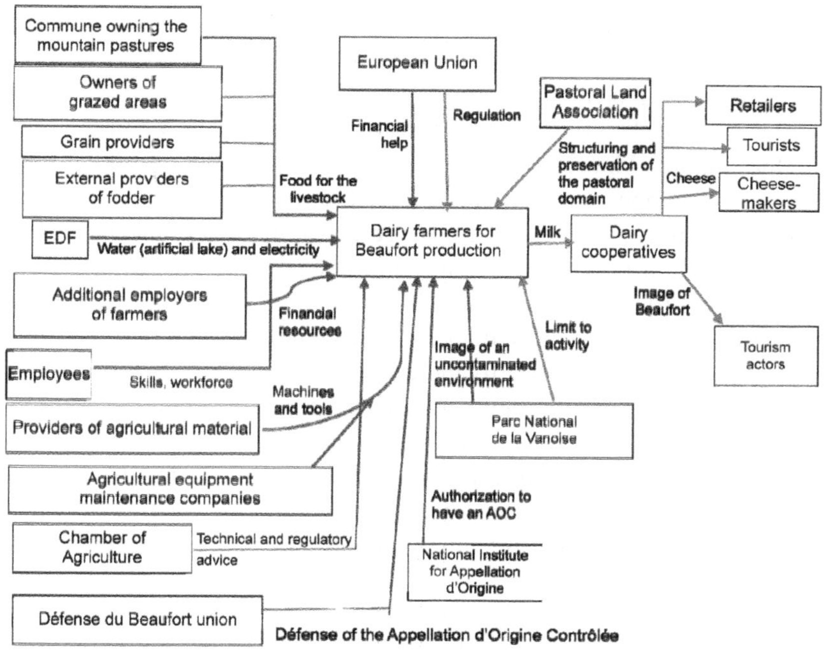

Figure 3.7. *Representation of the role of stakeholders in wealth creation: the case of Beaufort cattle breeders in Aussois. For a color version of this figure, see www.iste.co.uk/buclet/territorial.zip*

For each relationship between dairy farmers and other stakeholders, we interrogate the type of interaction and, therefore, the type of flow concerned with this interaction. The arrows in blue correspond to the flows to the farmers (flows corresponding to the activation of resources). The arrows in red correspond to actions that facilitate/regulate/limit farmers' actions. The green arrows correspond to the flows produced by the farmers and which can correspond to resources for other stakeholders (such as the fact that the presence of cows in the mountain pastures, coupled with the production of Beaufort, can constitute a resource that can be activated by tourism stakeholders).

This diagram echoes previous diagrams illustrating the mobilization of resources (stocks) and the circulation of flows within a system (and in interaction with the outside). In order to do this, it is proposed here to determine the stakeholder to be placed at the center of the system. The choice of this stakeholder very quickly depends on the geographical scale considered. Thus, in the Beaufort example, not on the scale of the commune of Aussois but, for example, on the scale of the Haute Maurienne, the relevant stakeholder is the Haute Maurienne cooperative. At this scale, the stakeholders involved are also quite different. We will find a certain number of stakeholders at the Haute Maurienne, or even the Maurienne or Savoie scale, playing a structuring role for the activity. They are in particular collective type stakeholders, capable of federating individual stakeholders (farmers and cooperatives) in order to increase the efficiency of coordination and, consequently, the wealth produced. If, on the scale of Aussois, we are able to identify the existence of an AFP (*Association Foncière Pastorale*), whose objective is to preserve and structure the pastoral domain through the adoption of collective rules between landowners and public financial support, many support structures are established on a larger scale.

In the case of livestock farming in the Maurienne, the importance given to the collective is reflected in a number of structures. This is the case of agricultural groups (GDA – Groupe de Développement de l'Agriculture – of Moyenne Maurienne and GDA of Haute Maurienne), associations which, since the end of the 1960s, "defend collective approaches that take into account all points of view. True centers of reflection and positioning, interfaces between the agricultural world and local authorities, the groups are in charge of raising local issues and conducting collective projects in the area"[3]. Similarly, the *Société d'Economie Alpestre* (SEA) of Savoie was founded with several objectives to consolidate the pastoral sector in relation to economic, environmental and tourism concerns, while the Beaufort sector is structured on a Savoie scale with the *Syndicat de défense du fromage de Beaufort*, which acts as a collective stakeholder on a national or European scale.

These last considerations show us that the metabolism of a territory is strongly influenced by collective considerations. In fact, this invites

3 Available at: http://www.agriculture-maurienne.com/service-aux-agriculteurs/les-gda-moyenne-maurienne-et-gida-haute-maurienne.html (accessed March 25, 2020).

us to continue our reflection on the questions to be asked concerning the relationships between stakeholders.

3.2.2.2. *Interests and motivations of the stakeholders*

Determining the reasons that shape the metabolism of an activity therefore requires an analysis of the motivations that lead individual and collective stakeholders to direct their actions in one direction rather than another. From this point of view, in the context of the work at and on Aussois, we also briefly sketched out what consideration certain concepts and methods could bring.

The question of motivation has been worked on in various fields (not only in psychology, quite logically, but also in economics or management sciences) and at different scales (economic organizations, political systems, individuals, etc.). What is important for us here is to take into account the great diversity of individuals who act in relation to other individuals, or within collectives. Anthropology seems to us, more than any other discipline, to have seriously investigated this question. Its disciplinary ambition is to think about:

> the general functioning of the social and cultural and to identify universal analytical categories capable of explaining both the diversity of human societies and the unity of the human race. (Kilani 2012, pp. 8–9)

It does so by trying as far as possible to avoid establishing universal rules of behavior from a local point of view (as the economy has done by imposing a situation constructed from certain postulates adopted at a given time in a given region of the world and by making it a "universal law" of human behavior). The work of Alain Caillé, to which we will return later, has been oriented toward this question of motivation, notably in response to the economic science that has established the figure of *homo oeconomicus* as a universal figure. Without going back over the details that led to the dogmatic imposition of this individual, driven solely by his or her egoism in the name of his or her individual freedom[4], it seems useful to recall that the considerations and theses developed by certain thinkers such as Adam Smith (1776) and Jeremy Bentham (1889) were established as impassable and universal truths, far beyond what these authors had in mind when they evoked the idea of the invisible hand (Adam Smith) or the principle of utility (Jeremy Bentham and then John Stuart Mill). A human society must certainly deal with egoism and the defense of the interests of individuals, but even in economic matters this is far from sufficient to explain the nature of interactions. Thus, from an anthropological perspective, the notion of needs, which is fundamental in order to deal with the creation of wealth, depends on the symbolic or social value of objects as well as on social relations. Similarly, the notion of scarcity is indeed a cultural construct (Sahlins 1973) and abundance depends on social rules that may or may not limit the desires of the members of a community. As Mondher Kilami reminds us, in primitive societies "there is a tendency to under-production and under-utilization of work capacities" (2012, p. 141), probably linked to the absence of a need to have more and more.

It is therefore important, in order to understand the current dynamics within the metabolism of wealth-creating activities, to analyze the motivations behind social interactions based on cultural knowledge of the territories. In this respect, interviews with the stakeholders involved in a wealth-creating subsystem must be conducted, in order to understand what the activity in question represents for them and the motivations behind their involvement. In our opinion, these motivations are of two kinds: motivations linked to exchanges with other stakeholders and socio-cultural motivations.

3.2.2.2.1. Motivations linked to exchanges between stakeholders

As we announced earlier, we take the position that we base this analysis of motivations on anthropology, particularly economic anthropology. Alain Caillé, at the heart of the anti-utilitarian movement in the social sciences, and therefore at odds with the vision

4 I would like to refer to Buclet (2011a), but there is no shortage of works on the subject, in particular, the excellent work by Christian Laval (2007).

of *homo oeconomicus*, offers us a reading grid of motivations that can explain our way of acting and especially of acting when interacting with others. This reading grid is summarized in Figure 3.8.

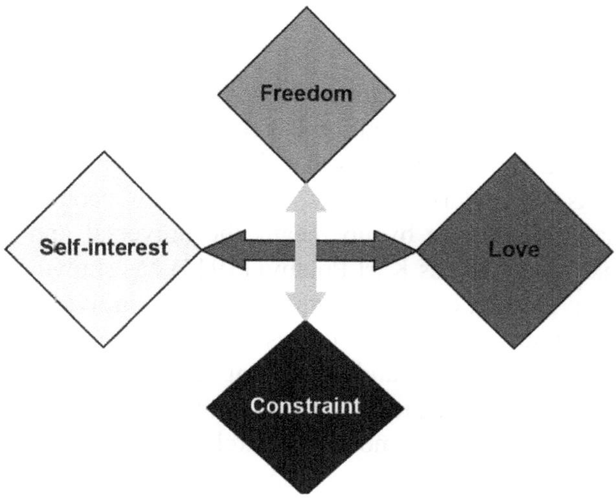

Figure 3.8. *Diagram of individual motivations in interaction phase (source: Caillé 2009). For a color version of this figure, see www.iste.co.uk/buclet/territorial.zip*

This grid highlights the existence of four main areas of motivation. In addition to self-interest, which economists believe is almost the only motivation for each of us, not only in today's world but also universally, other motivations are just as important. If the aspiration to individual freedom is compatible with the selfishness on which each of our actions is based, love and constraint have no place in this selfish vision. By love, we must understand sympathy for others, while by constraint we must understand that each of us accepts social constraints in order to integrate ourselves into a human community. Relationships between individuals are therefore driven by both interpersonal and social motivations.

Alain Caillé specifies that none of our actions are the result of a single source of motivation. Ambivalence prevails when we act. We are motivated both by self-interest and by sympathy for others, which is a source of internal tension (tension represented by the red arrow). Similarly, we aspire to be able to act freely, to free ourselves from the constraints of social life, but as a social animal we bend and respect the conventions of the society we live in. This second source of tension is represented by the green arrow connecting these two poles.

Regarding this concern, adopted by many anthropologists to find invariants beyond the socio-cultural specificities of the various human societies, Alain Caillé considers that this scheme has universal value. It allows us to understand why, even in the economic domain, commercial exchange is by no means the universal and impassable mode of exchange but, as Karl Polanyi (1944) has shown, only one of the possible modalities of exchange between individuals within a society. Sympathy for others drives us to express this sympathy through concrete gestures, and gift-giving is part of these gestures. In his various works, Jacques Godbout (1992, 2007) shows us that gift-giving is one of the modes of exchange between individuals, including in situations where purely selfish approaches or purely functionalist reasoning should prevail (as in the operating mode of organizations with an economic purpose, such as companies, where the emotional component of relationships is fundamental if we wish to explain the reasons for the behavior of employees and managers).

Gift-giving is also found as an expression of an individual's need for inclusion in a social group. Because of the "strange loop" (Godbout 2007) inherent in giving and giving back, we often give more than we have received and make others our debtors just as we are theirs. The social cement within a community is explained by the multiplication of gifts, which calls for reciprocity, but a reciprocity without equivalence (otherwise one would be in a market-type exchange), a reciprocity that does not aim to measure debts owed to one another.

3.2.2.2.2. The coupling between motivations and exchange modalities

It is for this reason that the motivational approach deserves to be coupled with Karl Polanyi's work on the modalities of primal exchange within a given human society. Let us recall that based on a large number of works and documents from human societies located at very diverse times and places, Karl Polanyi (1944) established a triptych characterizing the modalities of exchange with an economic purpose. This triptych proves, like Alain Caillé's work, to be a challenge to the economic dogma centered around *homo oeconomicus*. In addition to market exchange – understanding the exchange of goods on the basis of a price established from the interplay of supply and demand – this author highlights redistribution, operated by a central stakeholder who captures the wealth produced by a human community and then redistributes it to the members of the community, and reciprocity based on the extension of the gift between members of the community. A human society is characterized by the co-presence of these three modalities of exchange, to varying degrees. In an archaic society, the place of reciprocity is generally important, intertwined with redistribution. The place of commercial exchange is much more modest. In our society, market exchange is predominant, but depending on the country, redistribution also plays a rather important role (less so in the United States than in France, for example). As for reciprocity, although it has been totally ignored for a long period of time (which coincides roughly with the development of national accounting and the measurement of GDP based on a monetary measure of human activities), more recent works attempt to revalue its importance without measuring its weight in society. Indeed, this form of exchange, omnipresent in our relationships, including economic relationships (see Jacques Godbout, already cited), has the particularity of escaping the monetary measure that governs our society.

This is how a human community, on the scale of a State, as well as on a smaller scale, can be analyzed according to what emerges from reciprocity, redistribution or market exchange. This is what gave Serge-Christophe Kolm (1984) the idea of a triangle within which countries position themselves according to the weight of these three modes of exchange.

Thus, we have a double reading grid that can be applied to all the metabolic flows of a wealth-creating activity. To which exchange modality can we associate the circulation of a flow between two stakeholders? What motivates each of the stakeholders involved at the origin of the circulation of a flow, what motivations are intertwined? What tensions are perceptible between stakeholders and what real or latent conflicts can be identified?

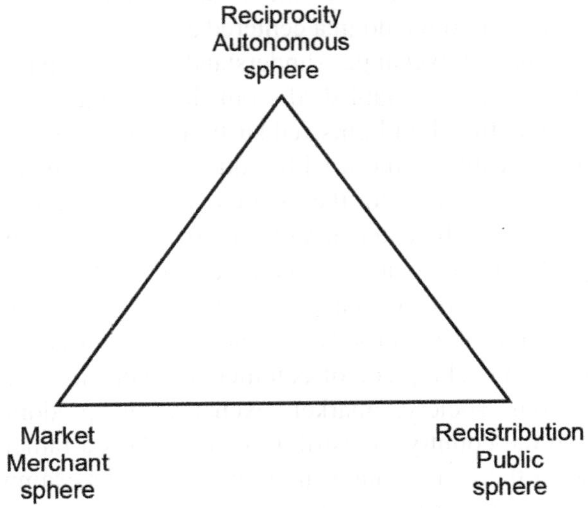

Figure 3.9. *The Kolm triangle (source: Van Parijs 1991)*

3.2.2.2.3. Socio-cultural motivations

Before turning to the last question raised just now, let us now look at motivations not at the level of interaction between individuals, but at those stemming from belonging to a community. Anthropologists tell us and show us how our behaviors can be different from one culture to another, as needs and desires diverge between civilizations. One need only refer to the famous works of Pierre Clastres (1974) or Marshall Sahlins (1973) to understand that the strange is in the other, in that all our reference points about what is or is not important are shaken, even overturned, by the way of thinking and acting of this other, of these other human communities. If, as Karl Polanyi (2008) has shown with regard to the interpretation of the status of tankarums in Mesopotamian cities, the first anthropologists often found it difficult not to explain the facts observed and the discourses collected other than through the prism of their own society, this remains one of the great challenges for any anthropologist immersed in a community that is strange to him or her. In particular, Karl Polanyi points out how utilitarianism used as a prism of behavioral analysis has distorted the observed reality in favor of a functionalist and economist vision. Similarly, Mondher Kilani reminds us that a great anthropologist like Bronislaw Malinowski seized the collected material in a utilitarian way. This is illustrated, for example, by his definition of culture as "society's organized response to basic or cultural needs" (Kilani 2012, p. 239), as if culture existed as a vehicle for an explicit purpose. We would only act out of need, whether that need is a physiological need or a need for social integration. Claude Lévi-Strauss, who raised this limit in the work of Bronislaw Malinowski, states: "to say that a society functions is a truism; but to say that everything in a society functions is an absurdity" (Lévi-Strauss 1958, p. 17). To this, Mondher Kilani adds that "it is indeed frequent in a culture that an institution does not fulfill any function or, on the contrary, fulfills several functions at the same time" (Kilani 2012, pp. 240–241).

Thus, any interrelation between stakeholders does not necessarily have a functional reason to exist. These cases are nevertheless likely to interest us, even though our approach is strongly based on a certain functionalism. This is because we consider that, even without explicit purposes, behavior has an effect on society, including indirectly on activities with an economic purpose.

Fieldwork based on qualitative interviews therefore seems necessary at this stage, in order to identify more precisely the motivations that make the various stakeholders involved in a wealth-creating activity act, whether these motivations are functional or not. If we take again the example of mountain agriculture, considered for several decades as a "survival" of tradition "destined to disappear under the effect of mountain modernization" (Kilani 2012, p. 281), we can see that its permanence is largely due to non-functional reasons. It is not uncommon to find that agricultural activity continues thanks to investments made possible by the dual activity of farmers. In other words, some of these people would earn a better living without the agriculture they practice. What motivates these people? Is it just habit and fear of change? Whether it is the work of Gérald Berthoud and Mondher Kilani (1984) on animal breeding in the Swiss lower Valais or the interviews conducted in 2019 in the Maurienne by Julia Grosinger in the framework of her thesis, the interviews give comparable results. At least some of the farmers (often dairy farmers in these examples) continue their activity because they feel attached to their way of life, and so continue to live in the village and participate in the collective life of the territory. Even though tourism is the main source of income for these farmers, they do not build their identity from tourism, which would push them to maintain agriculture "for the folklore", but conversely they use tourism as a tool to perpetuate the aspects of the traditional way of life that they wish to preserve.

We can see from this example that it is relevant not to confine oneself to individualistic and material motivations, but to have a framework that complements the collective reference points of the occupants of a territory. It is at this price that it is possible to insert the various interactions linked to a wealth-creating activity into the lifestyle of a territory taken as a whole. We shall return to some of the consequences of these considerations in Chapter 5, in terms of avenues of reflection for the future of territorial ecology. In the meantime, it should be remembered that the analysis of the circulation of flows between stakeholders involves both an analysis of the direct motivations in favor of the circulation of a flow and an analysis of the embedding of this circulation within a broader socio-cultural context.

3.2.2.3. *Territory: a land of agreements and conflicts*

Other aspects deserve to be taken into account in this work analyzing the relations between stakeholders, that of power games and the conflicts inherent to these relationships. Indeed, we cannot tip over into the naïve vision of a territory endowed with a unanimous vision and that orients its actions in a single direction. After having highlighted the stakeholders to be taken into account, while underlining the importance of power, legitimacy and interest, we have so far looked at the motivations related to the question of interest. It remains for us to look more closely at the questions of power and legitimacy, which are closely interrelated, even if there are forms of power that do not directly lie within forms of legitimacy. To do so, we will rely on the orders of worth that we owe to sociologists Luc Boltanski and Laurent Thévenot, who wrote a seminal work

(Boltanski and Thévenot 1991) on the questions of legitimacy of the discourses and actions of individuals as well as institutions.

Beyond the motivations of one and the other, there is a need to agree with other stakeholders in order to be able to act. A wealth-creating activity presupposes coordination efforts, implicit or explicit, which can be a source of tension between stakeholders. These tensions can be of two different kinds: a divergence in the the system of legitimacy within which to coordinate (Godard 1990) and a divergence within the system of legitimacy shared by the stakeholders.

3.2.2.3.1. Legitimacy and contestation of actions

We are not going to make an exhaustive presentation of the orders of worth, nor of the economy of conventions that follows, but we will limit ourselves to showing the interest of an approach based on these theoretical contributions and to referring the reader to the numerous scientific publications in the field. Let us recall that Luc Boltanski and Laurent Thévenot developed a conceptual framework based on "worlds", or representations of the world, which aims to establish the people, institutions and objects that matter most. Each system of legitimacy, or order, is therefore a system of values. There is a plurality of these representations of the world and each one is intended to be universal, which implies a conflict between them. Depending on situations and contexts, stakeholders refer to one of these systems of legitimacy in order to justify not only their point of view, but also the actions they undertake. Among these orders is the market order, in which the value of individuals, institutions and things is measured in money and which legitimizes the prevalence of market exchange over any other mode of exchange. The industrial order measures value according to the ability to produce as much as possible with the least amount of resources, and to be efficient, even in the optimization of production processes. It goes without saying that these two systems of legitimacy are at the heart of our society.

Other systems can also be found there, in various forms. Thus, the civic order measures value according to the public, egalitarian aspect. Common goods have a higher value than private goods, especially when the latter are for private use. Democracy is of great value within this order, as is any device that creates the conditions for equitable access to means of subsistence or resources. With the inspired order, what is important is the views of the world, whether these are held by religious leaders, political figures or intellectuals. Value is measured in the singularity of humans and objects, their integrity, and the gratuitousness of relationships and things. This value is not measured, it is decreed, signified by those who know and who impose this knowledge.

The domestic order places value on proximity, on belonging to the same family, understood in a broader sense, in the same village, or even the same territory. Memory is a fundamental means of attributing value to people, institutions and objects. Heritage, in its various meanings, is a fundamental reference point. The order of opinion values what is known or famous; what is entitled to media coverage. This value is measured by notoriety.

In addition to these six orders introduced by Luc Boltanski and Laurent Thévenot, other orders have emerged, which are however less unanimous. Among these is the interesting project-based order (Boltanski and Chiappello 1999), which measures value according to the ability of individuals, institutions and devices to produce a network and make a useful contribution to the process leading to the realization of a project. We see in this order the increased importance of management and a world of work where individuals must demonstrate their ability to work on projects. We must also mention the ecological order (Lafaye and Thévenot 1993), which gives value to nature and to the stakeholders and institutions that speak for it. In particular, nature which is valued is not confined to a private good; it is only marginally related to that which is already deemed valuable in market or domestic orders. Rather, it is a planetary good.

In the phase of interaction with others, a stakeholder refers to one of these worlds of justification in order to legitimize either their point of view or the action they will take. It frequently happens that individuals or institutions have to interact without sharing the same advice concerning the system of legitimacy to which reference should be made. In this case, there is controversy, which needs to be resolved from the top down if we want to increase society's capacity for coexisting. Compromises, arrangements that allow both sides to find common ground, are in this case necessary.

The analysis of these controversies is useful to understanding the dynamics of a territory, in that it allows us to understand each other's logic of action, the values they attribute to objects, nature (Godard 1990) and the relationships between stakeholders. In a territory, flows circulate between stakeholders, but it is important for us to understand on what basis, whether it is conflictual or consensual (which enables us to establish the more or less strong legitimacy of each other's action), and with what end expressed in terms of the value attributed to this or that aspect of the transaction. In other words, the use of the orders of worth model allows us to identify the intention of the stakeholders in relation to the flows that circulate. By identifying what the flows mean for the stakeholders, we can identify the stakeholders who are more successful in imposing their point of view on others. It is therefore also a representation of power games that emerges from this analysis.

3.2.2.3.2. From legitimacy to power

The metabolism of a territory is certainly the result of the intentions of the stakeholders, but even more so of the intentions of the dominant stakeholders, who manage to impose their point of view and, therefore, the values that should be privileged within an activity that creates wealth. Thus, in today's Western society, the largest stakeholders according to the values of the market order are more likely to impose their point of view than others. The financial means at their disposal (because the more one "weighs" monetarily and the more one is worth in this order) exceed the zone of influence of the markets. Lobbying, now considered legitimate after having been practiced in a hidden way, considerably increases the influence of these people and institutions in the legislative domain and in the decisions taken by the executives, who are supposed to act primarily according to the values of the civic order. Likewise, no one would even think seriously to run for the presidency of the United States without possessing substantial financial means. The same is true of notoriety, a determining value in the context of the order of opinion, and whose media coverage depends greatly on the financial means invested. Victory in elections seems to depend more and more on the notoriety acquired by those with superior financial means, to the detriment of the arguments specific to the civic order. In other words, fields of public decisions which, in a democratic society, are considered as belonging to the civic order, are largely confiscated by stakeholders, who have more weight in the market order, and which lead those stakeholders that are considered as big in the civic order to follow their interests of market order.

And what is true for the civic order is also true in the field of the industrial order. If experts and engineers are supposed to be the greatest within this system of legitimacy, if the most efficient productive processes are supposed to be privileged, we observe serious limits posed by the market order on the industrial order. Thus, the fact that efficiency is measured in terms of the use of resources according to their monetary value, and not their productive value, produces considerable bias. Industrial processes generate considerable pollution but are considered financially profitable and therefore continue to do so, failing to give any value to the environment (see, for example, Stéphane Héritier's (2007) article on oil shale exploitation in Alberta). Since the environment is not valued in monetary terms, its degradation counts for nothing in the market order. Faced with this state of affairs, environmental economists explain that a monetary value must be assigned to the environment so that it can be taken into account[5]. In other words, faced with the primacy of the market order in a growing number of domains where other orders and their ranges of importance would have their space to legitimize other ways of doing things, the only answer given to us by the economists consists of increasing the weight of market legitimacy in society, rather than in favoring a plurality of ways of thinking.

As we can see, an analysis of the discourses of legitimization of postures and acts allows us to highlight the power relations between stakeholders. We are thus likely to better understand the power games that guide the way in which a wealth-creating activity develops and is

5 See the article by Julien Milanesi (2010) for a critique of this monetarization of the environment.

structured, according to the weight of the stakeholders and their ability to make this weight legitimate.

This analysis by the orders of worth model applied to the metabolism of a wealth-creating activity allows us to address the question of who decides the circulation of flows, an important point for understanding territorial dynamics. This important point will be taken up again in Chapter 4 when we study territorial capabilities.

3.3. Conclusion

In writing this chapter we had the ambition to describe a new method of analysis in territorial ecology. The idea was born in 2013 in Aussois, during the thematic school already mentioned (see Buclet 2015b). For the time being, it has been mobilized only marginally, if we compare it with work in social ecology, work using the Eurostat method, or the determination of the environmental footprints of cities. The few works that have been carried out in the past (Herbelin 2018; CAP-Bioter project financed by Ademe, BOAT project financed by Ademe) or that are currently in progress (doctoral theses by Michela Bevione and Julia Grosinger at the University of Grenoble Alpes) are still too few in number for it to be possible to base the method on sufficiently established results. The approach is evolving as a result of this work, which has contributed and is still contributing to the clarification of certain methodological gray areas.

After having established the main guidelines of the method, and in particular insisting on both the importance of the immaterial part of the metabolism and the analysis of the intentions of the stakeholders behind each flow, we are going to advance on two particularly important methodological aspects: how to analyze the resilience capacity of a territory, especially in the face of global changes? How can we better take into account the interactions between human societies and their environment beyond questions of available resources? These two questions intersect, as we shall see, around the notion of capability that we owe to the Indian economist Amartya Sen.

4

Territorial Ecology and the Challenges of the Socio-ecological Transition: Metabolism, Capacity and Resilience

The previous chapter (see Chapter 3) has allowed us to explore several methodological issues as well as proposals to fill some of the gaps in the analysis of metabolism. Our main proposal, developed in Chapter 3, consisted of a realignment of the systems to be considered with the main wealth-creating human activities. This way of considering human societies, of dividing them into finalized subsystems, allowed us to propose extensions toward the analysis of why and how flows circulate, while adding immaterial flows, which, in a human society, are just as important as material flows.

Other issues deserve our attention, in connection with current events that are not immediate (the author is writing this book in the midst of the coronavirus crisis), but which for years should be at the heart of public policy beyond the rhetoric, namely, the necessary modification of our individual and collective behavior in the face of ecological upheavals. In Chapter 1, we addressed the question of the socio-ecological transition in order to position territorial ecology in relation to the institutional stance. Our remarks here are of a different nature. It is a question of seriously taking into account a territory, classically included in a broader institutional, social and economic context, and which must seriously consider its future, following the explosion of a world of certainties, environmental control and rationalization of all that surrounds us. In other words, it is the uncontrollable ecological context that needs to be taken into account, something that has not been done for decades in the name of development and well-being, considered rational based on what appears to be controllable and therefore rationalizable. This does not mean the abandonment of action or of the project, but the abandonment of ways of reasoning and acting based on the credo of absolute mastery of our environment. It is in this sense that we intend to talk of socio-ecological transition; a transition that more closely takes into account the co-evolution between nature and society, as well as the capacity of a human society to transform itself, in order to increase its capacity for resilience, taking into account this necessary, or rather inevitable, co-evolution.

For this reason, this chapter is dedicated both to the methodological question of a better consideration within the territorial metabolism of the co-evolutionary relations between nature and society, and to the question of how to think about the resilience of a territory in order to face the necessary transformations. These two questions will be dealt with together, starting with the concept of capabilities, which we owe to Indian Nobel Prize winner Amartya Sen. This concept, which is closely linked to those of freedom and justice, deserves a presentation before we discuss our two questions.

4.1. Capabilities

This part may seem superfluous, as Amartya Sen's work has been read, commented on and mobilized. However, it seems necessary to us to approach this concept, the interpretation that we have of it, in order to redirect it (which happens with any successful concept) and to introduce the notion of territorial capability.

4.1.1. *The rise of individualistic freedom*

As we have already seen in Chapter 3, freedom is one of the main motivations explaining the behavior of individuals, especially when interacting with others. However, freedom is not a fixed, irremovable concept to which human societies commonly refer. Before it had connotations to primarily individual freedom, the philosopher Roberto Esposito reminds us:

> that freedom, originally, has a community connotation. It is a power of connection, of aggregation, of pooling: even if it is more in the sense of belonging to a common root that grows and develops according to its internal law, than in the modern sense of "participation". (Esposito 2019, p. 121)

Before being the freedom to act within the community without taking into account its binding rules, and in particular its moralizing rules (a strong stance from Bernard de Mandeville's fable of the bees written in 1714, see in particular Mauroy (2011)), freedom originally linked to community was understood as the place "of plurality, difference, otherness" (Esposito 2019, p. 127). This is what makes Roberto Esposito – who wishes to set liberalism and communitarianism against one other as two facets of the rejection of that which binds a society – say that "freedom can only be 'common' – to each and every one, because it is nobody's own" (Esposito 2019, p. 127). Is it possible to be free alone? We are free only through others (Char 1983, p. 733), just as we exist only through the eyes of others.

Modern freedom was built in ignorance, or even the rejection of this conception of freedom. It must be recognized that in the context of de Mandeville, Helvetius, Hume, then Smith and Bentham, society as a community no longer had anything favorable to plurality, difference or otherness. As Jacques Godbout reminds us, modern individualism:

> liberates the individual from the community, from the family, from all those bonds that constrain and prevent him or her from freeing himself or herself and becoming a "true" individual, with no other obligations than those he or she has given himself or herself. (Godbout 1992, p. 108)

Yet the aspiration to this type of freedom emerges in a society where otherness is not encouraged, unlike in so-called "primitive" societies where the individual is encouraged to act in the community according to his or her capacities and therefore his or her singularity (as Pierre Clastres (1972), shows in particular). Feudal society does not grant the individual this freedom but condemns him or her to a role predetermined by his or her birth into one family rather than another. Freedom finds no place in this social structure, limiting behavior to a morality imbued with Christian virtues. The scientific and intellectual revolution is slowly emerging from this structure and a new freedom is being put forward; a freedom that evolves alongside virtues and justifies material well-being from this breakthrough. It is freedom against the community that imposes its diktat on individuals.

4.1.2. *A certain approach to freedom*

This intellectual movement, at the foundation of our society, found an important echo, particularly with the French Revolution, which constitutes a cornerstone and thus a period rich in fundamental landmarks to legitimize the new world. According to the 1789 Declaration of the Rights of Man and the Citizen, liberty consists of doing anything which does not harm others: thus, the exercise of the natural rights of each person only has those borders, which assures other members of the society the fruition of these same rights. These borders can be determined only by the law (Article 4 of the Declaration).

As for the law, it is no longer based on prior morality but on the expression of the general will. In this context, it is sufficient to respect the law in order to not hinder the freedom of others. If everyone respects the law and if justice punishes those who break it, society is supposed to be able to progress toward an increase in well-being, thanks to the individual initiatives of beings driven by their own selfishness (this is the famous (and already mentioned) invisible hand of Adam Smith).

We can, with Isaiah Berlin (1969), consider this form of freedom as negative freedom (or freedom from interference). We are free from what we are not prevented from doing, either by legislative framework or by individuals who break the law, the consequence of which being a suspension of freedom. But according to this author we must consider another form of freedom, positive freedom (or freedom as self-mastery), which implies that we are free when we are able to decide what to do with our existence. This form of freedom is more demanding than the first one, in that it is not enough to pass laws and make them be respected. It is necessary to develop collective action that empowers the members of a society to act according to their own objectives. For example, a society in which some members are malnourished even though there is no general food shortage is a society that does not defend an essential positive freedom. Similarly, a society in which some of its members can only eat thanks to schemes to help the poor, is also a society that fails to respect the positive freedom of its members.

The defence, or even the increase of individual liberties, is therefore based on collective decisions that go beyond legislative texts, and are translated into concrete results allowing each individual to be able to act in order to meet his or her own needs. Taking up the work of Isaiah Berlin, Amartya Sen (1999) introduces the notion of capability as the autonomy of individuals in their ability to find the means to meet their own goals in terms of well-being. But this autonomy has nothing to do with any kind of autarky. It is based above all on the structure of society, the nature of institutions and the goals they set for themselves. In other words, the autonomy of individuals depends closely on the society to which they belong. It is the social conditions that increase our potential for autonomy, our capabilities. It also makes sense to reason in terms of the capabilities developed by a society for the benefit of its members, since this is what will enable each individual to meet his or her own needs expressed through a personal life project.

4.1.3. *Individual capabilities or collective capabilities*

Many criticisms have been made of a theory that is considered excessively focused on the individual (Lessmann and Rauschmayer 2014). This led to the development of an evolution of the concept of capability and the introduction of collective capability. These criticisms seem well-founded in that Amartya Sen's approach still remains too close to the dominant economic paradigm, even though his *homo oeconomicus* is not only materialistic. However, given the fact that individual capabilities only exist through society, a broad debate is open as to the appropriateness of introducing these collective capabilities. Amartya Sen (2014) himself expressed his reservations about this opportunity, arguing that individual capabilities are the product of society. What meaning would collective capabilities have then? What would they add that is not already inscribed in the concept of capabilities? Peter Evans, the first to introduce this variant of capability, insists that collective action may not be essential for those already endowed with capabilities, but that it is essential for the less privileged (Evans 2002). This argument seems insufficient, again in view of the dependence of capabilities on the institutional mechanisms of human societies. But then, does this mean that all the capabilities of individuals should be put on the same level? For our part, we consider, on the contrary, that there is a need to distinguish between capabilities that, although they are part of a society, exist without the need for explicit collective action, and those that cannot emerge without specifically dedicated collective action. The literature that has developed in favor of collective capability (see in particular the seminal article by Solava S. Ibrahim (2006)), following the already cited article by Peter Evans, has placed great emphasis, at least in the early years, on the importance of social structures, especially for disadvantaged individuals and communities. More generally, individual capabilities do not really take into account the interaction of individuals within groups that generate non-existent capabilities for its members without collective action (Rauschmeyer *et al.* 2015). Collective capabilities are therefore different from individual capabilities not by nature, but because of the process that generates them. They are the result of the free and voluntary involvement of individuals in a collective process (Pelenc *et al.* 2015, p. 228), which implies that in order to understand whether a capability is individual or collective, it is necessary to analyze the process of its constitution.

4.1.3.1. *Links between capabilities, empowerment and commonalities*

In this sense, it seems that collective capabilities are singularly close to the work of Elinor Ostrom and the need to build individual empowerment at the level of human communities, i.e. the possibility for individuals to actively participate in the decision-making process concerning strategic aspects of life choices (Pelenc *et al.* 2015, p. 230)[1]. This also brings us closer to the notion of "commons", revisited by Elinor Ostrom in order to question, on the basis of field observations, the famous tragedy of commons highlighted by Garrett Hardin (1968) on a purely theoretical basis. The latter indeed considered that common goods left freely accessible and unregulated could only lead to ecological disasters. Thus, a meadow left free for grazing without any restriction would be overgrazed, with each farmer having an interest in putting as many animals as possible on it. According to Garrett Hardin, the privatization of common property and the introduction of access prices would be the solution to this problem.

There is no doubt that, without regulation, the overexploitation of water from a river or meadow is quite systematic and leads to the degradation of the common resource. Unregulated free access to a resource poses a problem. However, a resource with free access is not a common resource. Elinor Ostrom, winner of the Nobel Prize in Economics, has shown a number of examples of collective management of common goods that are no longer freely available. The use of these commons is closely regulated by more or less explicit rules; traditional rules in some cases, collectively decided and modified rules in other cases (Ostrom 2009). This bears a resemblance to the constitution of collective capabilities. The question then arises as to what differentiates them. Among the examples of the management of commons, it can be seen that the rules established can be particularly unequal, because they are based on the prior rights of families. In other words, the management of the commons is not systematically linked to questions of justice, contrary to the capability approach. Coupling the commons approach with the capability

1 Which Amartya Sen considers the agency.

approach makes it possible, on the one hand, to insist on the importance of institutional arrangements in the management of fundamental collective elements at the local level, and on the other hand, to pay more attention to "real and equitable access to the resource and thus to the most deprived and vulnerable with regard to this resource" (Fontaine 2016, p. 6) and to move from a solidarity often based on tradition to a democratically elaborated solidarity.

There is therefore a complementarity between these concepts that should be taken into account when analyzing the relationship between mobilized resources and the increase in capabilities at both the individual and collective levels.

4.1.3.2. *Resources and joint action*

Common action is found at the intersection between collective capabilities and the governance of common goods. Mobilizing a resource implies action. When this resource is managed collectively, it generates capabilities that are themselves collective. Beyond, or in addition, to the notion of what is "common", which refers mainly to objects, Pierre Dardot and Christian Laval (2015) are interested in the "common" as a political principle that produces common action. This "common" is made up of political principles and values from which the members of a community determine a way of acting and mobilizing the resources found in the implicit or explicit functioning of institutions. We will make the link between this "commonality" and the conventional regime in Chapter 5, in order to question the importance of changes in principles and values in the face of sustainability issues.

What interests us more specifically here is an emphasis on the link between a large number of resources and common action. In the case of the commune of Aussois, we were able to observe the existence of an irrigation system common to farmers (Buclet and Cerceau 2019). As observed in other municipalities of the Alps, but also in very different geographical contexts (Ingold 2008), common irrigation systems constitute one of the best studied commons. In many cases, and this is the case in Aussois, the plots are separated by canals, while metal hammers regulate the flow of water available to each plot. While the system is undeniably material, with the challenge of distributing a decisive flow for an activity that creates wealth and is itself structuring in the mountains, its effectiveness depends closely on the possibility of joint action, the product of collective deliberations between members of the community concerned. The result of these collective deliberations, of their being put into action within the community, "allows the 'water deposit' to become a 'water' resource, but also a 'fodder' resource" (Buclet and Cerceau 2019).

But a territory can also be crossed by common actions that go far beyond its geographical perimeter. Still in the case of the commune of Aussois, the joint action that led to the recognition of the Beaufort AOC (*appellation d'origine contrôlée* – controlled designation of origin) at the scale of the Beaufortain, Tarentaise and Maurienne regions in Savoie is of prime importance in the ability to generate the resource or, at least, to renew this resource. The modernization of agriculture, a major challenge for the European Economic Community aimed at food autonomy, has weakened traditional agriculture in less favorable contexts, such as in the mountains. It was soon considered obsolete and destined to disappear in the long term, before being reevaluated at various levels. The joint action that led, in the 1960s, to the establishment of the AOC is one of the many initiatives that constituted an effective response, not only for the agricultural world, but also more generally for mountain territories and the preservation of a certain number of cultural landmarks (Kilani 2012). The milk produced has a much higher value thanks to this common action, in that it incorporates not only proteins and other nutrients, but also identity resources incorporated into the designation. These identity resources would have little or no value without this common action, made possible thanks to guarantee bodies such as the INAO (*Institut national de l'origine et de la qualité*, formerly *Institut National des Appellations d'Origine*), itself subject to European rules.

The common action took place at two levels:

– The grouping of farmers in cooperatives: making it possible to pool production tools and produce a pressed cheese cooked according to a technique developed in the late Middle Ages in the Swiss Jura (Mélo 2015) and to create the possibility of exporting the cheese produced in the Alps. These cooperatives are, in this case, an indispensable intermediary for this type of production technique, which requires production in large volumes.

– The joint action of cooperatives: structuring themselves within the *Union des Producteurs de Beaufort*, which provides technical support to the farmers. This *Union des Producteurs de Beaufort* is integrated into the *Société alpestre de la Savoie*, which contributes to the concerted management of the department's mountain pastures. The latter participated in the creation of the Beaufort Defense Union, which interacts with the INAO to promote and defend the AOC, and with the CNAOL (*Conseil National des Appellations d'Origine Laitière*), which brings together all the Defense and Management Organizations (ODG) for French dairy AOCs.

This structuring enables the daily management of an intangible resource that guarantees traditional know-how, a source of wealth for the territories concerned, in particular to ensure that the individual practices of the farmers respect the established specifications. Joint action is based on the objective of creating and then managing a resource that cannot exist without collective action and the establishment of principles shared by the stakeholders concerned. It is also in the sense that it is judicious to speak of collective capabilities distinct from individual capabilities within a social context.

4.1.4. *Territory and capabilities*

The examples at the crossroads between collective and common capabilities regularly lead us to a territorial scale. Does this mean that collective capability is necessarily territorial? We cannot confirm this and this would require a debate with the scientific community working with the concept of collective capability. Can we talk about territorial capability in order to better establish the link between this scale of common action and capability? The first authors to take the plunge are France Loubet, Jean-Christophe Dissart and Benoît Lallau (2011). These authors aim to make the link between the territorial economy and territorial capability:

> The territory thus constitutes a "structure of living together" that influences opportunities at the individual level. (Loubet *et al.* 2011, p. 685)

This approach is consistent with Amartya Sen's point of view, which, contrary to the introduction of collective capability, defends the idea of the "context in which the individual evolves" (Loubet *et al.* 2011, p. 687). Thus, these authors propose the notion of territorial capabilities as being:

> individual capabilities "anchored" in the territory, stemming from the latter, and which can only be obtained at the scale of the territory. (Loubet *et al.* 2011, p. 687)

We thus keep with this meaning, in an individualistic perspective of access to endowments linked to a context, even if it is territorial, and of conversion of these endowments into capacities.

Our point of view is different. It is certainly not uninteresting to look at the territoriality of individual capabilities, but this is not enough to justify the introduction of a new notion. What we mean by territorial capability is more a "specific form of collective capability anchored in a territorial context" (Buclet *et al.* 2015, p. 113). It is a matter of taking into account a relevant scale in terms of community of common destiny for humankind and the capability to collectively design a common project. Moreover, as we have seen above with the question of commonalities, good ecological management is frequently at the center of the problems. The question of resources, of their sustainability in order to ensure the capabilities of future generations, deserves just as much consideration. It is therefore also a question of taking into account the ecological potential of the territory, the capabilities being linked to resources that should be activated within the territory. In this sense, we do not only consider capabilities at a given moment but also as a potential to be preserved, or even increased, for the years and decades to come.

We can therefore say that territorial capability is a situated collective capability, but that is not all. As a collective situated capability, it is articulated not only with the resources given and constructed, but also with the local positioning of stakeholders and proximities. In our opinion, what really distinguishes it from collective capability is the link that we wish to highlight with the territorial metabolism. A territory whose stakeholders are able to increase their capacity to produce well-being depends on the equilibrium of the metabolism, and on their autonomy in the choices to be made in order to preserve this equilibrium. Thus, the closed loop of substances such as nitrogen is the guarantee of increased sustainability, in that it reduces pollution (Billen *et al.* 2013). Nitrogen, phosphate and potassium deficiency undermines the ability of an area to feed itself, and the closed loop of flows is the best guarantee to avoid significant environmental imbalances. Territorial capability must take both of these issues into account, as well as the autonomy of the stakeholders of the territory concerned in terms of real choices to modify the circulation of these substances in their territory. Real choices, or positive freedom, because it is a bit reductive to say that farmers "only have to" change their agrarian practices. They probably cannot do so if they are under financial pressure imposed by intermediaries between them and consumers, or if their products, although of superior ecological quality, do not stand out from others on the market and therefore cannot be sold at a price that takes this quality into account.

With this example, we hope to convince the reader of the importance of this coupling between collective capabilities and territorial metabolism, justifying the use of the notion of territorial capacity.

4.2. Better integration of the relationship to the environment within the metabolism

Since the work on metabolism has been developing, we have seen various trends emerge and then consolidate. Some approaches focus on the circulation of a particular substance and, in this sense, are above all directed toward the question of the impact of this substance on environmental balances. While they provide fundamental knowledge, this work is rarely linked to territorial dynamics in the way that we have discussed. Other work on metabolism, notably inspired by the Eurostat method, highlights the materiality of society and emissions to the environment, but fails to illustrate the less quantifiable aspects of the relationship between human activities and the environment.

This is even more true for the method proposed in this book, based on the analysis of the metabolism of wealth-creating activities. The main criticism that can rightly be made is that interactions with the environment are reduced to mere questions of resources to be mobilized. Even if this is not entirely accurate (we must recall that among the products of wealth-creating subsystems are also emissions to the environment), this criticism deserves to be taken into account. It is part of the more general debate on the notion of socio-ecosystems. This notion is the result of the convergence between the analysis of social systems and the analysis of ecological systems, considered necessary by a growing number of researchers. We speak of socio-ecosystems, socio-ecological systems or coupled human and environmental systems (Le Page 2017). Based on the work already mentioned by Elinor Ostrom (2009), the structure of a socio-ecosystem has taken shape around four components: the resources, the systemic whole including these resources, the users and mobilizers of resources, and the system of governance.

Our challenge here is not to go over this notion here, but to recall how much our methodology can gain by getting closer to work on socio-ecosystems, in that it attempts to improve understanding of the links between nature and society. In particular, the issue of biodiversity has not been taken into account, even though it is a fundamental element of an analysis of territorial capacity, since biodiversity represents both a pool of mobilized resources and one of the essential elements for the sustainability of socio-ecosystems over time.

Before plunging into this socio-ecosystemic approach to wealth, let us note with Philippe Descola that an approach that is as close as possible to reducing the border between nature and culture represents efforts that:

> are condemned to remain vain as long as they are based on the premises of a dualistic cosmology, as long as they presume the existence of a universal nature that multiple cultures code for, or to which they adapt. (Descola 2005, p. 121)

Thus, we must be careful not to consider the socio-ecosystems approach as universal. It is nevertheless appropriate for the West and its "naturalist" ontology, as defined by Philippe Descola, an ontology to which our way of thinking is linked, despite our wish not to lock ourselves into it.

4.2.1. *Socio-ecosystems and nature's contribution to people*

One of the major methodological challenges is to reconcile the issues of wealth creation with the maintenance of ecological balance over time. The track envisaged is to mobilize the notion of territorial capacity in order to identify the articulations between stakeholders and resources favorable to this conciliation. For this, it is advisable to refine the work around the patterns of wealth-creating activities. The issue is to better situate the place of interaction with natural ecosystems in these schemes.

Not all natural ecosystems used for the creation of wealth are internal to the perimeter of a territory, whatever it may be. And it is of course important not to neglect the interactions between a territory and the rest of the world, from an ecological as well as from a social, cultural or other point of view, rather than purely from an economic one. In any case, whether it is a remote ecosystem or the natural ecosystem within the territorial system concerned, the relationships can be envisaged through the notion of nature's contribution to people (NCP), a notion that tends to replace ecosystem services, which is considered too economistic. The authors behind this proposal (Diaz *et al.* 2018) insist not only on the fact that NCPs are co-produced by nature and humans, but also that this co-production is understood differently depending on the cultural context. In other words, an element of nature does not inherently produce a service of a certain value, but it is the way a society interacts with that element (which is rarely isolated from its ecosystem) that generates a benefit. As for this benefit, let us keep in mind that the notion of NCP is not yet sufficiently symmetrical, since it is about the benefit for humans, whereas it would be conceivable to also take into account certain symbioses that are also favorable to a better state of an ecosystem (even if it is not possible to express the existence of preferences on the part of the natural ecosystem). Like the notion of a socio-ecosystem, the notion of NCP remains anchored in the naturalist ontology of the West, and it may be sensible to question its operationality in cultural contexts marked by the absence of dualism between nature and culture (Descola 2005).

If we accept this point of view, and if we accept that this work can only be coherent with a naturalistic vision of the world, we can then consider that it is a great conceptual step, in that it allows us to highlight the crucial importance of interactions between humans and other components of the biosphere, without *a priori* on the best way to interact, within the limits of the sustainability of the socio-ecosystem. Moreover, as this notion emphasizes cultural aspects, it becomes all the more sensible to not focus only on physical interactions. The stakes of nature–humanity interaction are also immaterial, or ideal, if only because of the place of the symbolic in the way humans look at what surrounds them. In a Western context, this symbolism is linked to the notions of heritage, landscape and identity, notions that are important in that we link them with the notion of resources.

4.2.2. *A socio-ecosystemic approach to wealth*

Let us examine once more the diagrams in Chapter 3, in particular Figure 3.1, which represents the basic architecture of the wealth-creating subsystems.

The question that arises is the place of NCP in this scheme. It leads us to consider the co-construction of resources on a territorial scale (including with inevitable geographical overruns), in accordance with this logic where resources are a social construct according to the literature on territorial economy (see Kébir and Crevoisier 2004). We will therefore not only consider that a resource is a socio-ecological construct, but also that the wealth-creating activity combines resources with each other in order to produce wealth.

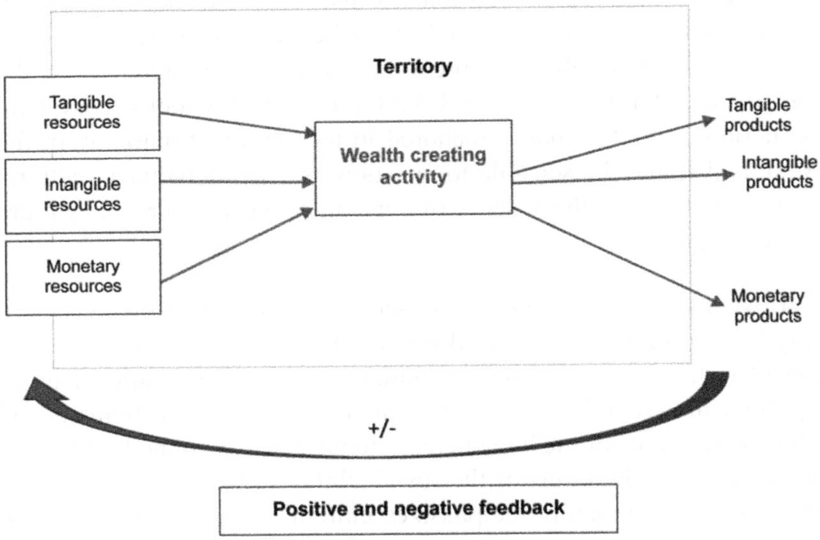

Figure 4.1. *Wealth-creating subsystem (source: Buclet et al. 2015). For a color version of this figure, see www.iste.co.uk/buclet/territorial.zip*

4.2.2.1. *Socio-ecological resources*

This makes us consider that at least part of the material and immaterial resources are socio-ecological resources. We are aware that the very notion of resource is highly linked with an anthropological point of view. It is necessary for us to claim, not as an asset, but as a fact, that our way of reasoning is inscribed in this naturalistic ontology which led us to consider the existence of a nature which is external to us (which is not universally considered, see again Philippe Descola for that) and that this nature represents a potential mobilized by humans. Even though we now know that this ontology has led us to reduce the planet's sustainability and that it is appropriate to think of our human society within a single biosphere (Buclet 2011b), it is intellectually difficult to detach ourselves from the way we are used to perceiving a certain type of link with what surrounds us. Also, without being able to subscribe to a meta-ontology that would embrace the various ontologies observable on the scale of humanity – a meta-ontology that is difficult to think about since ontologies are all designed to explain the world and are therefore not very compatible with each other – we will have to be satisfied with the worldview in which we have been immersed since our birth.

According to this vision, socio-ecological resources are a construct resulting from the will of stakeholders who intend to mobilize nature through knowledge, beliefs and techniques. As an example, mountain pastures constitute a socio-ecological resource – one that is unthinkable without the mobilization of an ecosystem, but a resource that does not exist as such without human intervention. The alpine ecosystem does not constitute an *a priori* resource. It is human societies that have modified this ecosystem according to a specific issue: the production of fodder for livestock. For this socio-ecological resource to contribute to the creation of wealth in a territory, it must be integrated into a wealth-creating subsystem within which it is combined with other resources. In the Maurienne – our illustration field around the production of AOC Beaufort cheese – this resource is coupled with another socio-ecological resource, the cattle that feed on this fodder and from which milk of a given quality is obtained. The two breeds of cattle that make it possible to produce Beaufort cheese, the Tarine and the Abondance, are themselves the result of interaction between humans and their surroundings, and not breeds found in the wild.

These socio-ecological resources are mobilized together with other resources, such as know-how in the production of pressed cooked cheese, a technique developed in Switzerland which then spread to the rest of the Alps in the 16th century (Mathieu 2000), as well as the development of common irrigation systems to increase the yield of grazed or mowed plots (as is the case in Aussois), as well as the culture of solidarity which for a long time has been decisive in difficult environments above a certain altitude.

Nature's contribution to people is thus fully in line with the identification of a large number of socio-ecological resources mobilized within the framework of a wealth-creating subsystem. There is also convergence between the proposed representation and the notion of bundles of ecosystem services (Crouzat *et al.* 2015). These bundles are the result of joint effects between the biophysical characteristics and the history of a territory. This notion challenges us on the need to not deal separately with the various ecological contributions of nature to human societies, neither from a spatial nor a temporal point of view (Crouzat *et al.* 2015, p. 1146). Thus, it is important to consider the articulation of socio-ecological resources among themselves within the same territorial system, and also to take into account the fact that the socio-ecological resources present are the result of multiple interactions that have taken place in the past.

The legacy of past interactions between humans and nature must therefore be taken into account. This means that the socio-ecological resources mobilized at a given moment must be considered as a legacy of past years, just as the potential of the socio-ecological resources of the future is partly shaped in the present, by the way we shape what surrounds us and influence climatic dynamics, etc. This temporal thickness varies with the speed of change in the environment, which is slower in cold environments than in equatorial zones.

4.2.2.2. *Socio-ecological products*

The same reasoning as for resources should guide us, again in conjunction with Figure 4.1, with regard to the products of wealth-creating activity. At least some of these products have consequences for both society and nature. For example, the fact that the production of Beaufort cheese requires alpine pastures has in return different consequences. Mountain pastures constitute a rich ecosystem from the point of view of the diversity of the flora and are very different from the one that would exist in the event of agricultural abandonment and re-wildening. There is a co-production with the natural environment, an environment that is frequently highlighted as a "postcard image" and therefore as a valuable landscape. This heritage landscape is a co-product of mountain livestock farming. It is also an immaterial socio-ecological resource, for the benefit of other wealth-creating activities, whether tourism or residential activity.

Socio-ecological products are therefore likely to constitute the resources of other activities, thus producing closer links between activities on the scale of a territory, particularly through their relationship with natural environments. Nature's contribution to people is thus made as a tangle of co-production of wealth (the wealth that a landscape constitutes as a contribution to well-being, for example), a large part of which is or will be remobilized as a resource of the territorial system.

However, since the products of wealth-creating activities are considered in a broad sense, not as all the elements that humans intend to produce, but as all the output flows, we have to take into account the production of flows that disrupt natural ecosystems (greenhouse gas emissions, sulfur dioxide emissions, polluted water discharges, etc.). This is also a case of socio-ecological production weakening the subsystems that create wealth, or even the entire territorial system and beyond in the short and, more often, medium to long term. This socio-ecological production, considered as negative by the societies concerned, is likely to weaken territorial capability in the more or less long term, whereas the socio-ecological co-production seen previously also deserves to be analyzed from the point of view of its effect on territorial capability.

4.3. Territorial capability and resilience: a means to tackle transition issues

Taking into account socio-ecological resources and products allows us, much more than a simple classical approach by resources, to integrate the long-term consequences for the society of its own functioning. The consequences for nature have no meaning in themselves, since nature can, according to the dominant ontology in the West, be self-sufficient. It is the consequences for nature that are important to us, including, more recently, those that have an impact on other living species to which we now attribute rights (starting with the Universal Declaration of Animal Rights in 1978).

A dynamic understanding of the capability of a territory requires an analysis of the socio-ecological consequences of past practices and the traces of these practices that we observe in the present when we study the metabolism of wealth-creating activities. Beyond the aspects of governance and the capability of the stakeholders of a territory to decide their fate, it is also important to understand whether past and present choices affect individual capabilities in the future. In other words, in order to be sustainable in the long term, territorial projects have to face the question of vulnerability, which is itself closely associated with the question of the resilience of the territorial system.

4.3.1. *Vulnerability and resilience*

The articulation between vulnerability and resilience is closely linked, in the scientific literature, to the notion of socio-ecosystems. According to W. Neil Adger, resilience refers to the intensity of a disturbance that can be absorbed without having to radically change the state of a system (Adger 2006, pp. 268–269). A resilient system is able to reorganize and adapt to new circumstances (Adger 2006, pp. 268–269). Vulnerability is somehow the negative side of resilience. It is measured by the stress to which a system is exposed, its sensitivity to that stress and its ability to adapt.

Coming from different scientific disciplines, these two notions overlap, even merge. As Janssen (2006) reminds us, resilience comes from ecology (after having been first developed in physics and also quickly been adopted by psychology (Dauphiné and Provitolo 2007, p. 116)). Vulnerability, however, is rooted in work on natural hazards and the issue of poverty (Janssen 2006, p. 238). Nevertheless, links should be made between these terms of different origin, beyond the fact that a more resilient system is of course less vulnerable.

Vulnerability in particular “integrates two other components: exposure to hazard and resilience” (Dauphiné and Provitolo 2007, p. 121).

Several authors emphasize the importance of adaptive capacity to reduce vulnerability. Another concept was also discussed. According to Smit and Wandel (2006), the concept of adaptation is itself first mobilized by ecologists before anthropologists fully understand it. Michael J. O’Brien and T.D. Holland define a process of adaptation as a process in which human communities add new ways of adapting to the environment to their cultural repertoire (1992, p. 37). Other authors define adaptation as a decision-making process and a set of actions undertaken in order to cope with changes or disruptions in the socio-ecosystem, without leading to significant changes in the system (Nelson *et al.* 2007, p. 397). According to all these authors, adaptation is always based on the idea of moving as little as possible from the point of equilibrium prior to the disturbances experienced. On the contrary, the resilience approach relies more on the idea that the natural state of a system is constantly changing (Nelson *et al.* 2007, p. 398).

Dauphiné and Provitolo remind us that systemic resilience (like that of a colony of ants) is directly proportional to the self-organization of a system (2007, p. 117). Faced with a disaster, it is a matter of accepting its inevitability and doing everything possible to reduce its impact, a strategy that seems to be increasingly necessary in the face of global change. For a society, one of the strategies consists of reducing single-activity territories and encouraging diversification of activities. Nelson *et al.* (2007, p. 409) further insist that the existence of rights and duties shared between stakeholders, as well as the decentralization of decision-making, are elements favorable to the resilience of a territorial system.

These considerations reinforce the idea of coupling issues of resilience with those of capability. Self-organization, decentralization and the sharing of responsibilities go hand in hand with an increase in the autonomy of individuals in their life choices. It seems important to us to take into account the capacity for the resilience of a society at the scale of individuals’ living space, which is very close to our definition of territorial capability. But what would be the characteristics of a resilient territory? In addition to the fundamentals of its ability to anticipate shocks and to cushion their negative effects, as well as its ability to adapt, we add its ability to learn, to innovate in order to preserve and improve the way it functions (Buclet and Donsimoni 2018). All these characteristics conducive to resilience contribute to the construction and deepening of a collective consciousness (Rabourdin 2012), a common to be mobilized to achieve a dynamic balance, insofar as a resilient territory is a territory in motion. Considered on a territorial scale, resilience is a setting in motion, a perpetual search for a dynamic balance between paradoxical characteristics and contrary processes: short term and long term, local scale and globalization, redundancy and efficiency, tailor-made and ready-to-wear, autonomy and dependence (Villar and David 2014).

4.3.2. *Capability and resilience of territorial systems*

Trade-offs are necessary between possible targets, in order to be able to respond both to sudden shocks (such as the one faced by many countries in this year, 2020, marked by fear of a virus) and to progressive shocks (climate change). If these shocks affect the metabolism of a territory, public policies are designed to produce a dynamic response capacity and resilience of the territorial system. In order to do this, public policies should be based on the articulation between the analysis of metabolism, which characterizes it in terms of both present and past flows, and the analysis of territorial capability, which looks at the potential for action of the stakeholders in the territory and their autonomy in terms of choices with regard to this potential.

Figure 4.2 schematizes the articulation between metabolism, territorial capabilities and resilience at the territorial system level.

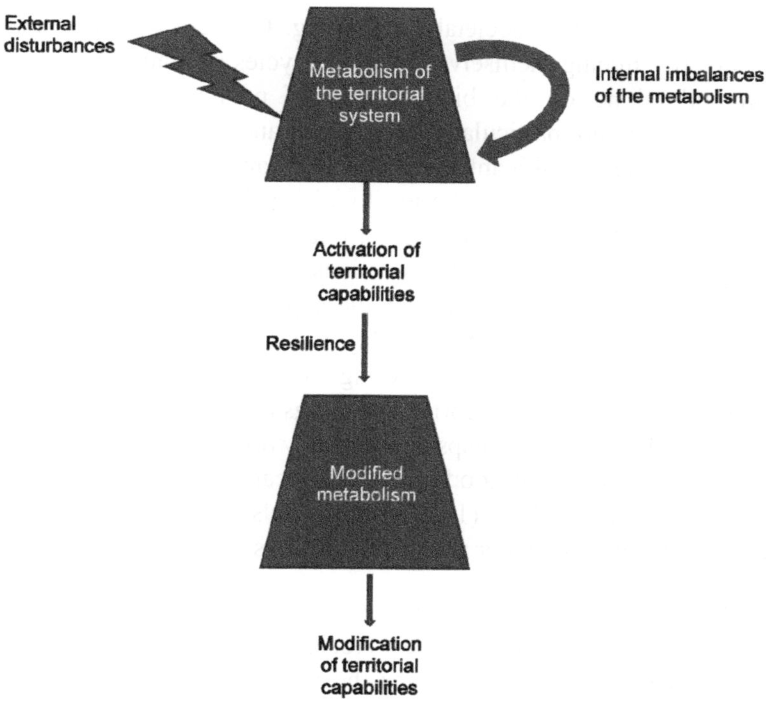

Figure 4.2. *Metabolism and capacity of a territorial system (source: Buclet and Donsimoni 2018)*

Except in unlikely cases in our society, the metabolism of a territorial system generates its own imbalances. A balanced metabolism would be based on an important internal looping of flows and on an equivalence for the main chemical substances between outgoing and incoming flows. We are far from this today in the industrialized world (in some cases we are almost at the antipodes), while societies considered archaic or primitive have persisted through time, in a strong symbiosis with their environment, both from a material and symbolic point of view (Descola 2005). From an internal point of view, these societies were balanced in that they were not based on any boundary between humans and non-humans (animals and plants are distinguished from humans by body shape and not by mind according to animist ontology (Descola 2005, pp. 183–184)) and therefore integrated many more variants that we would define as ecological (to use the words of our society) into their societal functioning. Conversely, industrialized societies (by freeing themselves from the cycles of matter, by ignoring them and taking on the biblical idea of nature at the service of humankind), do not particularly need disturbances coming from outside to regularly face significant metabolic imbalances. From this point of view, we can consider that the territorial capabilities and resilience of primitive societies are particularly strong, much more so than in our society. Of course, with a few exceptions, humans in the industrialized world would not aspire to live in these primitive societies, but it is important to remember that capability is not about owning more and having more goods, but about fostering the ability of individuals to live their lives without feeling deprived of things or opportunities that others enjoy[2]. In other words, perhaps because they do not know anything else, such as the profusion of goods arriving by cargo ship (see the myth of cargo ships by Baudrillard (1970)), individuals in these societies do not seem to feel any spiritual or material emptiness.

What about disturbances coming from the outside? It all depends on what we mean by "the outside". According to our conception of the Western world, the outside means the non-human: our environment. We would then have to refer to a tsunami, the coronavirus epidemic or climate change as disturbing elements coming from the outside. These are elements that come from nature, according to naturalistic ontology, and that we must learn to master. But not all societies consider these events as being external to society. Moreover, the concept of a socio-ecosystem also tends to no longer consider "natural" events as being external to the system to which we belong, i.e. the biosphere. Thus, the exteriority of these phenomena does not seem satisfactory to us.

But what does "external disturbances" mean? The territorial system being spatially delimited and separated from other territorial systems,

2 In this sense, let us recall again the work of Marshall Sahlins (1973) on the Aboriginal Australians, whose openly provocative book title (*Stone Age Economics*) is nevertheless telling (the title goes even further in French: *Age de pierre, âge d'abondance*, meaning Stone age, age of plenty). In these Aboriginal societies, individuals do not have a sense of lack, both from a material or spiritual point of view, in a surprising contrast to the feeling marked for the citizens of the industrialized world (see the numerous works inspired by the concept of *anomie* by Emile Durkheim).

we will consider as external what comes from other territorial systems, linked to actions undertaken by humans (and therefore including any environmental disturbances linked to this), whether with the intention of having an influence on our territorial system or not. However, here too we need to distinguish between two cases. The first is the influence on each other of territories based on the same conventional regime (Buclet 2011a) and, it goes without saying, the same ontology of relationship to each other (Descola 2005). The second, which we will examine first in order to focus on the first case, which is more relevant to territorial ecology, concerns the influence on each other of territories based on distinct conventional regimes and relationship ontologies. In other words, interactions between populations of different civilizations.

From this point of view, so-called "primitive societies", as soon as confronted with so-called "modern" and subsequently "industrialized" societies, proved to be particularly vulnerable. The foundations of their territorial capabilities, i.e. coexistence with other living species, have been strongly undermined by the encounter with human organizations based on the exploitation of natural resources. Majid Rahnema (2003), who conducted a survey of Aboriginal Canadians, shows the fragility of these societies in the face of the domination of the "white man", as does Serge-Christophe Kolm (1984) about the Pintubis tribe in Australia. But here we are not dealing with interterritorial interactions, but rather with intercivilizational interactions.

As mentioned above, the external disturbances of interest for our discussion are less radical disturbances, within the same civilization and with a large number of shared conventional reference points. Here again, these interactions are of two kinds: interactions with other territories, whether they are close or not, and interactions between the territory and stakeholders and institutions on a larger scale (national, international, global for a-territorialized stakeholders). In today's dominant conventional regime, the most important disruptions come from stakeholders and institutions with a wider range of actions than the scale of a single territory. Whether they are financial or political institutions, mass distribution or production groups, their influence is geographically wider and their weight on the dominant conventional regime is also greater. Beyond the ecological imbalances inherent in any territorial system enclosed within the dominant conventional regime – which we will discuss at greater length in Chapter 5 – it is therefore the actions carried out by these stakeholders acting on a large scale, sometimes planetary, that constitute the main sources of disturbance of the territorial metabolism. These actions either change the rules, or influence the possibility of financing territorial projects, or take decisions that have an impact on productive activities or consumption within a territory.

The challenge highlighted by the diagram is that of the capacity of mobilization that the stakeholders of the territory make from the potential of the territorial capabilities. On this mobilization depends the resilience of the system and the capacity of evolution made necessary by the disturbances.

4.3.3. *Beaufort cheese: between vulnerability and resilience*

It seems important to us at this stage to use the example of agriculture in the Maurienne valley, in Savoie, France, to illustrate these issues of vulnerability and resilience and thus highlight the relationship between metabolism and capabilities. More specifically, we will come back to the production of Beaufort cheese.

4.3.3.1. *Resources and products related to Beaufort cheese*

In Figure 4.3, we have not represented the flows with arrows, as is usually done in the case of metabolisms. It was important for us to represent the main resources mobilized by the production of Beaufort cheese and the main products resulting from this activity. This is not an exhaustive representation of the resources and products, it only shows the main ones. The resources as well as the products can be material (red), immaterial (blue), ecological (green) or financial (yellow). But above all, as we wish to show that most of them are a cross between material and immaterial, between material and ecological, between immaterial and ecological aspects. These last two crossings in particular represent the socio-ecological resources that should be preserved by an environmentally sustainable policy.

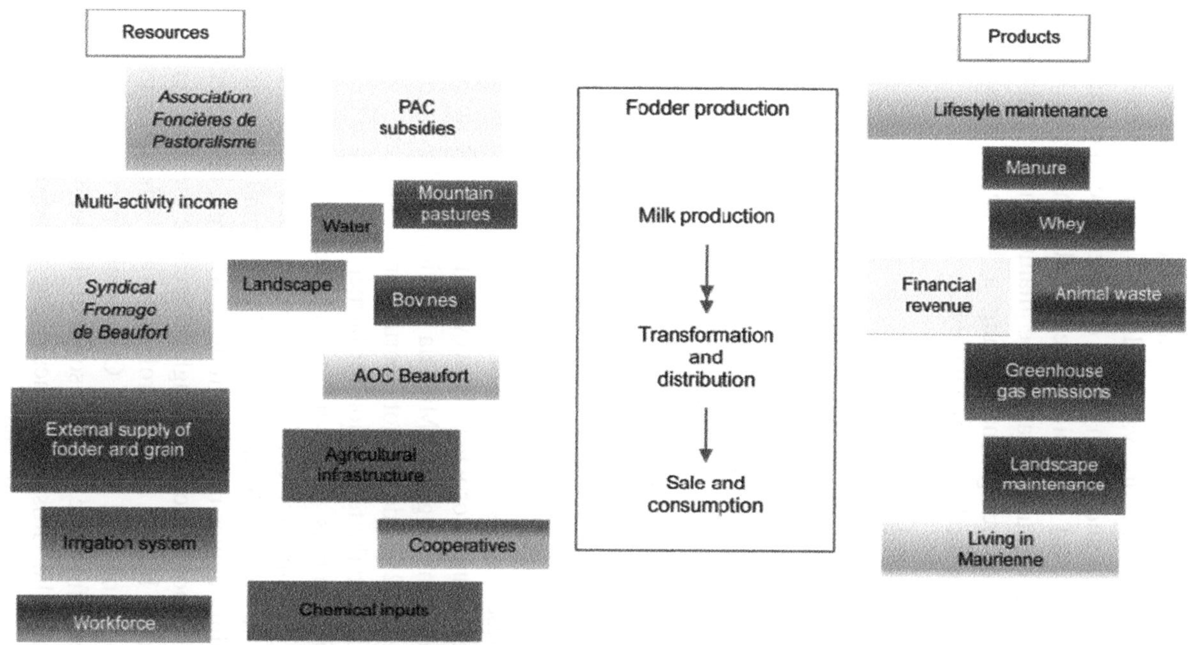

Figure 4.3. *Main resources and products related to the wealth-producing activity of Beaufort cheese. For a color version of this figure, see www.iste.co.uk/buclet/territorial.zip*

On the product side, besides the financial income (linked to the sale of cheese), there are: undesirable socio-ecological products (greenhouse gases, animal excrement, etc.); socio-ecological products which, if they are valorized, have positive repercussions (whey, manure, etc.) and which would otherwise be likely to degrade the ecological state of the soil and especially the water; products which enrich the resources (landscape maintenance); and products which increase well-being (maintaining a lifestyle, possibility of continuing to live in the Maurienne valley).

4.3.3.2. *Internal vulnerabilities*

Once the framework of what is circulating has been established, it is important to take stock of both the vulnerabilities of the activity (and relate it to the vulnerabilities at the scale of the whole territory) and the resilience generated by the mobilization of territorial capabilities. These vulnerabilities may be induced by metabolic imbalances within the system or by external disturbances, produced either by other territories or at a broader scale of the socio-ecosystem. At the internal level, imbalances may result from excessive exploitation of socio-ecological resources, mainly water and mountain pastures, or from an excess of chemical inputs resulting in water pollution. These excesses can result in greenhouse gas emissions, animal dejecta, manure or whey. However, it should be noted that the metabolic work in progress (Michela Bevione's thesis, due to be defended in 2021) does not particularly indicate significant imbalances, despite the temptation for farmers to increase the pressure (using chemical inputs or by increasing the size of the mountain pastures) on the environment in order to produce more feed. There would therefore be no particular vulnerability from this point of view. The vulnerability comes rather from the workforce, either because the remuneration is not attractive enough or because the new generations are not attracted by the farmer profession and what this implies in terms of constraints and lifestyle. Conversely, part of the territorial capability stems from the fact that some of the people interviewed (in the context of Julia Grosinger's thesis, due to be defended in 2021) appreciate this profession and the associated way of life. This interview-based survey work indicates that motivation in terms of monetary income is rather secondary. If this motivation was crucial, it is highly likely that most farmers would have decided to change activity, or even region, in order to increase their chances of increasing their cash income.

On another scale, however, it is possible to identify a vulnerability of the activity. Work on the effect of climate change on the Alpine world and in particular on the evolution of snow precipitation indicates that the latter will tend to decrease, at least below 2,500 meters, during the 21st century (Ménegoz *et al.* 2020). Knowing that winter sports resorts tend to accumulate snow stocks from one year to the next in order to reduce climatic hazards, we can make the rather probable hypothesis of an increased scarcity of water resources. This is true for the lower and middle Maurienne where the altitude is not high enough to benefit from significant snow cover in the decades to come. It is also true for Haute Maurienne whose climate is similar to that of the Southern Alps, which will in future be subject to more droughts. There is therefore a vulnerability that is likely to weaken territorial capacities and which the territory must take into account, a vulnerability to which it must react in a logic of increased resilience.

In connection with the question of snow cover, the question of the sustainability of an important resource for farmers also arises, namely, the possibility of combining their activity with a winter activity in winter sports resorts. Although according to the work carried out by Caroline Rivier (2019), this income from multi-activity is only about 10% of the average income received by farmers[3], this represents an increase in territorial capacity that should not be overlooked. This pluriactivity is linked to winter tourism, itself related to the vulnerability of the territory to current and future changes in the snow regime in the Maurienne. We will return below to the link between agriculture and tourism. It should be noted, however, that the question of internal or external vulnerability to the territory is not simple, as the intertwining of activities and between these and the outside world is so important.

---

3 This report is considered confidential and, unfortunately, we cannot disclose the main results.

The last aspect related to internal vulnerabilities is the link between farmers and cooperatives. The surveys of farmers conducted by Julia Grosinger as part of her thesis indicate both membership in the AOC Beaufort and recognition of the decisive importance of this appellation for their activity. This is as much due to the present advantages of the appellation as to the history that gave rise to its birth. It should be remembered that following the creation of the *Union des Producteurs de Beaufort*, the Beaufortin, Val d'Arly, Tarentaise and Maurienne regions obtained the AOC in 1968 (Faure 1999) and put a stop to what seemed at the time to be the inevitable disappearance of mountain breeding. The consequences of obtaining the AOC are very important, we will come back to this point concerning the disturbances coming from outside. What we must note here is nevertheless the vulnerability of this common action, in that some of the farmers who were questioned mentioned the possibility of selling their milk at the same price or even at a higher price to the mass distributors, while other farmers would like lower constraints included in the specifications of the AOC. These trends, if they were to expand, would constitute a potential threat to one of the pillars of the Maurienne valley's territorial capabilities, the AOC coupled with the operation of cooperatives.

4.3.3.3. *External vulnerabilities*

As noted above, external disturbances can destabilize a territorial system and test its territorial capabilities. External disruptions can mainly concern the external supply of animal feed, subsidies from the Common Agricultural Policy (CAP), the questioning of the Beaufort AOC, and the decline of tourism.

With regard to the external supply of feed, remember that it is limited to 25% of the daily intake of cows when they are productive, and there are no signs of any difficulties in the coming years. However, having to rely on a high level of local feed is a significant element and a resilience factor, provided that climate change does not seriously affect water availability.

The CAP has long been a part of the landscape. Its existence and the fact that it concerns the whole agricultural world is proof of the relatively low autonomy of European agriculture. The very fact that an entire sector, as strategic as the food sector, cannot do without subsidies calls into question the model of an increased opening of the economy in all areas, under the aegis of the WTO (World Trade Organization) and the will of the European Commission. An estimate made on the production of Beaufort in the Maurienne, within the framework of Michela Bevione's thesis, indicates that without European subsidies the price of Beaufort would have to be 35–45% higher in order to cover all costs (and the milk would have to be sold 55% more expensive to the cooperatives). This is not an exception since, according to Luc Smessaert, vice-president of the FNSEA (*Fédération nationale des syndicats d'exploitants agricoles* – the French National Federation of Agricultural Holders' Unions), CAP subsidies represent 47% on average of farmers' income[4], but this raises questions about the sustainability of this support, set up in 1966 and regularly contested by countries that are betting on a continuous increase in the productivity of a totally industrialized agriculture.

This point must be put in parallel with the dependence of the Tarine and Abondance breeders on the price of milk. We have indicated here one of the strong points of the territorial system, linked to the increased value of the production thanks to the AOC. Thus, in 2012, the purchase price of milk by cooperatives was almost double the national market price (Buclet *et al.* 2015). This advantage becomes even more important with the abandonment by Europe in 2015 of milk quotas and, consequently, increased pressure on the price of milk, due to the resulting increase in supply. Some industrialized farmers are increasing their herds (especially in Germany, Ireland and the Netherlands (Dubois 2015)), while others are doing everything possible to increase productivity per cow. This forward thrust helps to lower the unit cost of milk production. However, this is to the detriment of farmers. According to a German study (in a country considered to be at the forefront of modernity in the sector), in 2017 the average price paid to the producer covered only 76% of fixed

4 Available at: https://www.lemonde.fr/economie/article/2018/05/31/les-agriculteurs-francais-restent-tres-dependants-des-aides-de-la-politique-agricole-commune_5307535_3234.html (accessed June 10, 2020).

costs[5]! Therefore, despite the vulnerability that the dependence of farmers on European policy represents, the territorial capability of the territories concerned by Beaufort cheese is undoubtedly better than that of other milk producers in Europe. As the current trend is rather the increase in the value attributed by a growing number of consumers to the AOC, we consider that in its current state the questioning of the Beaufort AOC does not seem to be very credible.

The last point of vulnerability of the activity toward the outside is linked to the tourist activity. Indeed, a substantial part of the territorial system's capability relies on the sale of Beaufort at the local level, with cooperatives having their own stores and Beaufort being found in any food store in the valley. If this can be considered a strength of Beaufort, in that the producers and cooperatives are strongly autonomous with regard to a mass distribution system often denounced for its practices (back margins and other means of pressure[6]), then it becomes a point of weakness as soon as the tourist presence becomes rare. In particular, the effects of climate change on snow cover (Ménegoz *et al.* 2020) are likely to call into question a large part of winter tourism, which represents a decisive flow (about half of the tourist flow but above all two-thirds of spending in trade, crafts and catering; Buclet *et al.* 2015, p. 136). Thus, what represents a strong point of territorial capability is likely to turn into a weak point, given the global changes. Resilience here therefore requires a renewal of territorial capabilities, either by strengthening the distribution channels for Beaufort cheese while maintaining control, or by a strategy of evolution of forms of tourism in the Maurienne valley.

5 Available at: http://www.web-agri.fr/actualite-agricole/economie-social/article/en-2017-24-des-couts-de-production-n-ont-pas-ete-couverts-par-les-prix-payes-1142-135768.html (accessed March 14, 2018).

6 Available at: https://www.decision-achats.fr/Thematique/fournisseurs-1235/Breves/Les-entreprises-alimentaires-denoncent-pression-exercee-grande-distribution-250773.htm (accessed June 11, 2020). In this article dated February 11, 2015, the journalist recalls that four major group purchasing organizations concentrate 93% of food purchases. This represents an inordinate power of negotiation on prices and on decisions to sell or not to sell a product.

4.3.3.4. *A qualitative approach to be completed with a quantitative approach*

Although we have mobilized some figures, our analysis of the vulnerability of the wealth creation activity around Beaufort cheese is incomplete as it stands. Indeed, the legitimate question of the quantification of the territorial capabilities arises. This is one of the challenges in connection with the capability approach. Amartya Sen has taken the question on a societal scale by considering basic capabilities such as access to healthcare or access to school, aspects that are quantifiable and often even statistically quantified. At the territorial level, the question is more difficult. In addition to the aspects of governance and, above all, autonomy in terms of the choices to be made (autonomy linked both to the quality produced and to the operation of cooperatives), monetary aspects are also important. We have already indicated the share of CAP subsidies in the income of farmers, as well as the share of pluriactivity. A more complete work, already mentioned (Rivier 2019), has been carried out in order to understand the level of the main expenses of the farmers, as well as the composition of monetary income. This type of economic information, which is fairly classic, has its place in an analysis of territorial capabilities.

Other quantifiable aspects are the predictions of climate change models on a regional scale, as for example with the Adamont method, which allows us to obtain scenarios in terms of precipitation (which may or may not be snowy) at the level of the Western Alps and on a local scale, compatible, for example, with the Maurienne valley scale[7]. The development of climate change scenarios on a larger scale (and therefore on a more local level) is important, as is the development of the RAM (regional atmospheric model)[8], which allows us to refine our understanding of the changes underway on a scale of only a few kilometers. It is this data that should be mobilized, coupled in particular with work in ecology on the effects of these climate changes on vegetation.

7 Available at: https://opensource.umr-cnrm.fr/projects/adamont/wiki (accessed June 10, 2020).

8 Available at: https://mar.cnrs.fr/ (accessed June 10, 2020).

The work to refine the understanding of the capabilities of a territory is therefore at the interface of many disciplines and combines qualitative and quantitative aspects. In our opinion, it is not a question of arriving at a capability indicator. In a logic where this approach is likely to help the stakeholders of a territory to reflect on their future, to define a territorial project, it seems to us especially important to highlight the main points of vulnerability against which the territory is well armed or, on the contrary, in potential difficulty, in order to encourage a collective reflection on the ways to make the territorial system evolve and the articulation between wealth-creating activities.

4.4. Conclusion

The reader will have understood, we hope, that the approach we propose based on territorial capability still deserves methodological advances. Faced with the uncertainties brought about by global changes, as well as more sudden crises (such as the one experienced in 2020 with Covid-19), the capacity for resilience is today considered to be a research issue and a challenge for society. Our approach based on territorial capacity seems to us capable of contributing elements to the reflection on the need to make our society evolve toward greater sustainability. It is still a work in progress, theses are in progress and other research work, applied to territories that would be of interest, will be necessary in order to build with the stakeholders of these territories the way to elaborate a territorial capability approach. It will be difficult to make methodological progress without working with territories that are questioning their future and that wish to combine action and reflection and, therefore, also take the time to reflect.

The territorial capability approach, however interesting it may seem, should not, however, make us forget that a territory is part of a larger society, that modes of governance, the way economic activities are conceived, the way culture is given a place and, more generally, the values put forward, cannot be analyzed on this scale without making a link with the rest of the world. Change in the face of global challenges is undoubtedly located at the territorial level, but obviously in articulation with the broader levels of the large human community, and even the large community of living beings. Therefore, in the last chapter of this book (see Chapter 5), we will address the issue of innovation at the territorial scale in relation to the question of principles and values that can promote the sustainability of territories.

chapter of this book (see Chapter 3), we will address the issue of innovation at the territorial scale in relation to the concept of principles and values that can promote the sustainability of territory.

5

Principles for Innovative Territorial Projects

This final chapter aims to link the methodological aspects that structured the previous chapters with the question of change, the transformation of society in the face of global challenges. Indeed, the approach of territorial ecology focuses on the local aspects of socio-ecological issues, without any particular method to ensure measurement of the impact of these local aspects, not only on the rest of the planet, but also on other territories. As far as the effects on the rest of the planet are concerned, when it comes to non-spatialized issues, such as greenhouse gas emissions, analysis by metabolism is sufficient. On the other hand, global consequences with a spatialized impact (e.g. the disappearance of certain species in another part of the world) or consequences in other territories (e.g. the social consequences in the State of São Paulo in Brazil of the import of orange juice by the inhabitants of the Maurienne) remain absent from the metabolic balance sheet.

One solution would be to generalize the metabolism approach to the entire planet but at a territorial scale. This solution, which would be intellectually very attractive, would have the advantage of providing a global vision of both the relationships between territories and the imbalances between them. It would be possible to know whether the imbalances of certain territories compensate each other or, on the contrary, accumulate with a deleterious global effect on the socio-ecological balances of the planet. It would also be possible, beyond planetary imbalances, to identify the social and economic consequences of the metabolism of a territory in other parts of the world.

However, this solution is impractical. The work necessary for the development of territorial metabolism is not insignificant. It requires huge efforts and it is illusory to imagine that any territory in the world has both the means and the desire to embark on this diagnostic exercise; and even if this could be the case, it is just as naïve to imagine that the territories would agree to have the results of their metabolism circulating freely throughout the world.

However, it is important not to give up on understanding the effects of interactions between a territory and the rest of the world, both on a global scale and in terms of the privileged interrelationships between territories. In order to do so, we believe it is important to determine an alternative strategy that will allow us not to give up on understanding the interrelationships between different parts of the world.

Another solution, which would allow good control of flows and knowledge of them, would involve withdrawing from exchanges and favoring all that is local. Nevertheless, it seems preferable to us to pursue other avenues. We have already highlighted the dangers of autarchy (Buclet 2011a), the perverse effects of the confinement of individuals at several levels: the circulation of knowledge is essential to its maintenance; the circulation of human production makes it possible, under certain conditions of fair exchange, to avoid hostility toward others and xenophobia; and the circulation of humans makes it possible to confront others, to confront cultural differences and favors the questioning of certainties. While the territorial question remains a major issue at the local level – a crucial issue given our civilization has cut itself off from the world, has "gotten rid of territory [and] produced sites confined to a role of technical support for the economic system" (Magnaghi 2012, p. 11) – the excess of localism constitutes a counter-movement, potentially producing a large number of disadvantages. Alberto Magnaghi (2012, p. 14) rightly states that "a world without places is a world uninhabitable for man". In this affirmation, we see both the return of the necessary territoriality of the everyday and of the project, and also the importance of the plurality of places that humans can walk through in order to be enriched by the diversity they encounter. Our contemporary society and its search for financial efficiency tend to sweep away the differences between places, and even though we may be anchored in one place, this increasingly resembles other places, erasing the plurality that creates diversity.

The need for transformation of society lies between these two benchmarks and it is in light of these that we will consider the question of innovation. Why innovation and with respect to which watchword? While everyone refers to transition, whether ecological, energy or other (see Chapter 1), we prefer the notion of transformation. This need for transformation questions innovation. In a logic of transition, incremental innovations have their place: we improve the energy efficiency of a given process, we consume fewer resources to travel, to feed ourselves and to keep ourselves warm using the same vectors. Transition does not necessarily call for change, but for a change of state that may only be in degree and not in nature. In a logic of transformation, radical innovation, or disruption, is necessary. It forces us to question the meaning of what we do, the way we conceive of responses to our needs, and even our way of conceiving the very notion of need. The innovation we wish to speak of is therefore not conceived as tool that enables companies to differentiate themselves from the competition, to build competitive advantage, but as a means of transforming society.

If we associate the objectives of sustainability with the issues of territorialization and innovation, the question that arises and that drives us, concerns the capacity of local stakeholders to develop a transformative and sustainable territorial project. If territorial metabolism can be used by these stakeholders not only to understand where environmental vulnerabilities lie, but also to identify the territorial capabilities from which to act, we wish to show that it is relevant to couple it with a principles-based approach. To do this, we will build on the concept of conventional regime to present the main conventional reference points that structure our current society. We will then revisit the conventional principles likely to promote individual and collective behaviors in the direction of sustainability, before providing a certain number of elements for reflection on how to proceed, in an attempt to build sustainability.

5.1. Conventional regimes and principles

A human, just like a group of humans who never face otherness will be able to live in the comfort of certainty. It is when we meet others that we come up against different thoughts and it is when we leave our social group that we discover other ways of living. Without meeting others, we tend to think that our way of living and thinking is natural, that there is no other way. In recent centuries, travelers, followed by ethnologists who later became anthropologists, have written stories and then scientific works that astonish us about the diversity of ways of living in human communities. For a long time this diversity was perceived as a difference between more or less advanced civilizations, then, from the 1950s onward, as a difference in development. It was a difference of degree and not a question of alternatives. More recent works have broken away from this perspective, inspired by 19th century social evolutionism and by the exclusively Western view of the rest of the world. In Gilbert Rist's book, *The History of Development*, (2001), he strongly questions this vision of human societies developing according to a logic of progress. Other authors such as Michael Singleton refuse "to call development by anything other than its proper name: Westernization" (Singleton 2010). The acceptance of differences, no longer as transitory phenomena of progress toward a developed humanity, is now making its way and is being enriched by work on ontological differences, following the example of the work of Philippe Descola (2005), already cited on the ontologies orchestrating the relationships between humans and what surrounds them.

5.1.1. *Conventions*

Sociologists, as well as some economists, have, in their own way, questioned the diversity of individual and collective behavior and what this can imply in our relationship to the world. In line with Luc Boltanski and Laurent Thévenot (1991), an interdisciplinary French school of conventions was developed, which emphasizes, in particular, the plurality of modes of coordination between stakeholders, in line with the work of John Maynard Keynes. In an uncertain situation, this illustrious economist tells us, it is better to behave according to the dominant beliefs within a society (Keynes 1936), and, notably, according to the conventions in force.

A convention can only be considered as such from the moment when, in a given situation, no behavior naturally imposes itself. Should I drive on the left or right side of the road? Should I shake hands with people who come to me or should I lean my chest forward with my arms at my sides? Should I trust my interlocutor's word or should I draw up a contract that includes all of his or her obligations? Neither of these choices is obvious, and yet each of us, immersed in a socio-cultural reality in a given place and time, will have little or no hesitation as to the attitude to adopt. Conventions make it easier for us to deal with the uncertainty of the behavior of other members of a human community, since we can anticipate the behavior of others by assuming that they will conform to established social conventions and that they will similarly assume how we will behave.

Our behavior in society is therefore oriented daily by the dominant conventions within the society we are integrated into (or the society we wish to be integrated into). However, we are not always aware of this fact. It would be psychologically exhausting to be permanently in a state of reflection as to what to do in each circumstance; not to mention that, in some cases, this would be particularly ineffective, as Pierre Livet and Laurent Thévenot (1994) have shown, in terms of the ability to act in a coordinated manner[1]. In fact, we follow many conventions without being aware of them and this makes life in society much easier. We can even say that the most effective conventions are those we are not even aware of, those we follow as a "natural" way of doing things. This is the apparent paradox of the most established conventions: no one even realizes anymore that they are just conventions. They have become "naturalized".

---

1 Just imagine what the traffic would be like if one had to stop to confer with any other motorists passing by, to decide who is passing on which side.

When we talk about conventions here, we are talking about implicit or explicit rules of behavior at the inter-individual level. The conventions are numerous and activated in a decentralized way. And the more they are activated, the more they are reinforced, in a logic of self-reinforcement, highlighted in particular by Robert Sugden (1989), who considers conventions to be evolutionarily stable strategies in a context where there are at least two possible strategies. Individuals have an interest in adopting these strategies because they gain more from them than if they did not follow them and, in so doing, contribute to reinforcing the functioning of society, which is built on these numerous conventions. People who wish to follow minority conventions encounter difficulties (a feeling of exclusion, less effective in their actions and permanent misunderstandings with the population following prevailing conventions).

5.1.2. *Conventional values and reference points*

In addition to the conventions discussed above, those at the inter-individual level (micro level), societies also rely on conventions of a different nature and level. The members of a society do not only have concerns about efficiency in coordination, but also preferences in terms of values. There are two complementary ways of dealing with this issue from the point of view of conventions. The first is based on the model of Luc Boltanski and Laurent Thévenot (1991) previously cited and presented in Chapter 3. According to this model, values are found within worlds of legitimacy (orders of worth). The second is based on the existence of principles that guide our behavior and the way we represent the world.

5.1.2.1. *Legitimate orders*

The orders of worth model of Luc Boltanski and Laurent Thévenot aims to illustrate the plurality of representations of the world through the importance attributed to objects, people and institutions. Although each order has its own coherence and is thought of as a system of universal legitimacy, in a society these orders are not exclusive of one other. Individuals or groups of stakeholders mobilize them according to circumstances. A chartered accountant who is professionally immersed in a world legitimized by the industrial order, where everything must be predictable, planned, optimized and measured, does not behave (at least for those around them) in the same way with their family or friends. If they do, it is not said that the legitimacy they enjoy professionally is as great in their private life.

Depending on human societies, and according to the times, the legitimacy of certain categories of people, behaviors or objects evolves with the beliefs and principles on which each society is based. To clarify, let us recall the contempt that Greek citizens openly displayed toward any person getting rich through commercial exchange and let us compare this value judgment with the legitimacy in our society of commercial exchange and the stakeholders who claim to be so. We will come back to this point when it comes to deconstructing the unsustainability of our society.

5.1.2.2. *Conventional reference points*

We therefore attribute importance to things and people according to a system of legitimacy that we consider valid in a given situation. But how do we determine the relevance of using one system of legitimacy rather than another? What is it that makes market exchange degrading in some civilizations and a source of greatness in others (for example, among Phoenicians, contemporaries of the Greeks) or that public interest takes precedence in some countries and largely takes second place to private interests in others?

Every human community needs reference points to which each member can refer in order to guide their actions. To do this, communities adopt principles that their members agree to respect. We will not go into the details of how principles are adopted, as the process can be very diverse. We note two points that seem important to us:

– these principles are conventional in that they do not naturally impose themselves on any human community. Their adoption is not based on natural processes. They are the product of living together, are generally long-lasting and are part of the cultural traits of a community;

– beyond principles, reference points can be explicit or implicit. Explicitness can be very strong, as in a nation's constitution, or implicit, as in the benchmarks that everyone follows without even referring to any formal obligation.

The degree of implicitness varies in importance. In most cases, if those concerned are asked why they follow a certain principle, they are able to give a reason (it is customary, for example). In other cases, the benchmark, as such, is not even recognized, even though it has a determining influence on the behavior of community members. For the latter, it may be principles, or what we have called myths. More generally, they are conventional reference points that shape our individual and collective behavior in a very strong way. We will return briefly below to this category of conventional reference points with far-reaching consequences, as we deciphered and presented them a few years ago based on the myths that structure modern society (Buclet 2011a).

5.1.3. *Conventional regimes*

Before returning to these conventional reference points, let us consider the notion of a conventional regime. The simplest way to do this is to recall the definition we have already given:

> A conventional regime corresponds to a particular moment, at a certain time and in a definite space, when a human society is behaving according to a certain number of principles, which orient individual's behaviours within that society. (Buclet and Lazarevic 2015)

The regime is the third level of the conventional structure of a society, the first being that of interactional conventions and the second that of reference points. From this set emerges a regime within which we act and think in a certain way, based on certain beliefs. A regime is all the more stable when no member of the society questions its foundations. A society that is strong at a given moment is a society that is certain of its values, of the vision of the world that it carries within it. In these cases, it is possible to consider that this society is in a phase of stable regime. However, as we have seen in historical work on socio-ecological regimes, these phases of apparent stability often conceal less visible movements that are likely to weaken the regime (Barles 2005; Barles *et al.* 2015). The stability of the conventional regime is relative and depends on its capacity to absorb changes, without questioning its reference points. It is therefore a dynamic stability and not an immutable equilibrium state. This dynamic stability depends on the capacity of the regime to cope with the tests of legitimacy to which the reference points are subjected. In the face of hazards (a massive epidemic, to take an example at random), just as much as in the face of ongoing structural changes (changes in climatic conditions, for example), a dynamically stable regime absorbs shocks if community members are able to reinterpret the changes in light of conventional reference points, to give them meaning in action in a way that is both effective and in line with prevailing beliefs. This aspect seems fundamental to us in the dialectic between sustainability and innovation, to which we will return later. In fact, the innovation that is put forward so much, in order to take the path of socio-ecological transition, does not have the same meaning depending on whether it is part of the dominant conventional regime (we will come back to this in a moment) or it breaks with it. Nor is it evaluated in the same way, depending on whether the value attributed to it is part of the dominant conventional regime or of other conventional regimes, whether these are in the minority or in the process of becoming so.

5.2. From the dominant conventional regime to a more sustainable one

There is so much literature, both scientific and popular, on the problems of sustainability in our modern society that it would be illusory, and even tedious, to go through it all. In previous works, we have even contributed to this literature in order to distinguish a certain number of conventional reference points that characterize the dominant conventional regime, the one that locks us into a universe of beliefs and habits that we are struggling to get rid of. In the latter half of this chapter, the objective is to return to a certain number of conventional reference points that constitute the foundations of the dominant conventional regime on a global scale, before presenting certain principles likely to transform the situation and promote sustainable individual and collective behavior.

5.2.1. *The basis of the dominant conventional regime*

The reference points that interest us are not the product of an evolution explicitly pursued by present-day societies, but of a certain way of thinking about the world. While referring the reader to past works (Buclet 2011a) for a detailed explanation of the myths that structure our thinking and constitute formidable conventional reference points, we wish to highlight certain characteristics of the way of thinking about the world that leads to these myths.

5.2.1.1. *From progress to "hubris": the representation of Western civilization*

Our way of explaining the world and our way of living it are inseparable, as Philippe Descola (2005) shows expertly in his already quoted work on our relationship to what surrounds us. Our way of understanding the world today allows us to act by considering that what surrounds us is ours, that we can use it and appropriate it without having to ask anyone but other humans (or even not in some cases). And who should we ask since our vision of the world does not leave room for other thinking and soulful beings with whom we should talk, negotiate or trade? Alone in the face of nature, and moreover considering ourselves as the only ones to give meaning to it, we strive, through science, to discover all its mechanisms. The division of the world into more and more elementary bricks has given us unprecedented knowledge of the possible combinations between the ingredients such as chemical elements, then elementary particles. From there, the appropriation of what surrounds us is only the precondition for a massive transformation of our environment. This is what brought us into the Anthropocene era (Crutzen 2006), a new geological era in which humans are the main producers of background effects.

This increasingly intense transformation is the result of the scientific revolution which, starting with the Renaissance, led Westerners to no longer be content with mythological tales to explain the world. This has contributed to the emergence of a disenchanted world, to use Max Weber's expression, a world that would be without myths, that is to say a world whose foundation would no longer rest on a non-rational part of reality. The fight against the irrational has allowed Western societies to think that nature can be controlled for the well-being of humans. The most important step in the control of our environment was taken in England in the 18th century with the beginning of the Industrial Revolution, a source of undeniable progress in the capacity to produce more, as Adam Smith (1776) enthusiastically witnessed. Of course, this is an oversimplification of the historical process, and we are not unaware that a historian such as Jean Gimpel mentions a first industrial revolution in the Middle Ages (Gimpel 1975). No doubt the West or other civilizations have in the past shown inventiveness with the intention of multiplying the energy that can be mobilized. The fact remains that what Jean Gimpel describes (improvement of techniques concerning draft animals, exploitation of wind or hydraulic power, etc.) is more a matter of a more efficient appropriation of nature than a radical transformation of it. The mobilization of science, the progressive generalization of its principles in society, shifts a large part of the legitimacy of the stakeholders and objects of an order of worth, inspired by an afterlife, toward an industrial order governed by experts. We produce more and more, faster and faster, and will look for what we need to produce further and further away, whether it is at the other end of the world, under the Earth's crust or on the seabed. If Adam Smith and Thomas Malthus were worried about the limits of resources and therefore economic growth, it was only because they were giving a major share to agriculture and our ability to feed a limited population. Indeed, during the 19th century and especially in the first half of the 20th century, prospects were shaken by progress in all areas of materiality and the discovery of seemingly endless deposits, whether oil, metals, phosphate or uranium. It is this feeling of having broken the limits that had hitherto existed between happiness and humans that intoxicated the latter and led them – led us – to a state of *hubris* from which we must now emerge, or else collapse. Western civilization has thought of itself as a society of superhumans in a world to be dominated, whether it be other humans, animals, plants or even the chemical elements. New myths have emerged, even though Western civilization had decreed the end, which led Hassan Zaoual to say that "every society is based on myths. Man without myth is a myth" (Zaoual 2006, p. 115).

5.2.1.2. *The myths of a society without limits*

Several myths arise from this society of progress, myths that are important to us in that they represent conventional reference points whose influence is still felt strongly today, despite recurring criticism from parts of society. These myths are never made explicit as such, but are reflected in the discourses legitimizing the functioning of our society and seem to constitute unsurpassable societal objectives for a large part of its members, especially the dominant stakeholders, whether they be political leaders, large corporations, intellectuals or even influential researchers and academics.

These myths fall essentially into two categories, as we have shown before and as demonstrated in the book *Le territoire entre liberté et durabilité* (Buclet 2011a): founding myths of economic thought and myths linked to progress.

The myths linked to economic thought refer to authors considered to be pioneers of the economic sciences, and whose thought has been taken up, transformed and regurgitated in such a way that it does not question the march and even the course of an industrial, mercantile, capitalist and globalized society, to quote Anthony Giddens (1990). Let us recall here that the myth of the invisible hand, which we owe to Adam Smith (1776), has been absorbed by many economists and other thinkers as if it were a mechanism adapted to all human interaction. Thus, from selfishness and individual interest, pursued without limits, would arise an increase in general well-being. However, what is true for the butcher or brewer who in fact, in 18th-century Europe, did not exercise these professions so much for the well-being of their customers as for their own interest (although there are surely many other reasons that escape economic reasoning, such as the sense of belonging to a corporation, to a filiation, of feeling useful, etc.), is probably not true for all the interactions that make up a society. But it does not matter. Many economists have not bothered with such details, even if it means betraying the thinking of Adam Smith, author also of *The Theory of Moral Sentiments*, worth recalling, in which the idea of a multitude of motivations, beyond mere personal interest is defended.

What is important for these economists is to defend the idea of the supremacy of market mechanisms over any other method of coordination; and, in addition, the universality of this supremacy, to the point of turning them into "economic laws", real conventions that cannot be overcome within our society, even though they have long been contested by many stakeholders, including economists – one need only think of Karl Polanyi in the 1940s with his seminal work, "The Great Transformation". For this is what we are talking about, an ideology of economic liberalism, free trade and market supremacy that is first imposing itself in the West, then gradually throughout the rest of the world, even though a number of works demonstrate the limits, or the lack of proof of this supremacy. For example, the economic historian Paul Bairoch, with the help of extremely well-documented work, manages to show that free trade is not intrinsically superior to protectionism any more than the opposite, but that everything depends on the balance of power between countries (Bairoch 1993).

This ideology is enriched by founding myths, based on a number of "founding fathers of the economy" (in addition to Adam Smith, we can mention, Jeremy Bentham, whose thinking has been largely twisted, or David Ricardo, the latter more in line with liberal ideology), as well as on the idea of progress without limits. In this second order of ideas, the structuring myths of the dominant conventional regime are as follows:

– the myth of the more equal the better, which makes us consider that the more we produce, the more we possess, and the more we increase individual and collective well-being;

– the myth of the mastery of space-time, a strong tendency of a society that never ceases to want to go faster, to produce faster, to change everything faster (according to an absurd logic of change for change's sake, as Pierre-André Taguieff (2001) has shown), to move faster to the point of wanting to reduce the world to a network of increasingly similar points (homogenization of the places that "count") that can be reached and that are ideally connected instantaneously (hence, most recently, the "absolute necessity" to develop 5G as quickly as possible);

– the myth of control of the biosphere, which consists of no longer enduring the events imposed on us by the functioning of the planet, but understanding them in order to predict, contain or even modify them; and, consequently, no longer adapting our activities to natural constraints, but bypassing them, or even ignoring them in our rhythms of life[2].

Whether individually within society, or collectively, the movement of our daily lives is marked by these myths, to which we must add the myths of an ideologized economic science that has instituted so-called universal laws (among which is the consecration of economies of scale as the inescapable rule of efficiency). We impose on ourselves a rhythm of life that fewer and fewer people can withstand, a rhythm both in movement and in permanent change. Beyond the phenomenon of malaise, which is the subject of a vast literature (malaise at work in sociology, malaise and its electoral consequences in political science, malaise in daily life in psychology, malaise at school in education sciences, etc.), the feeling of the emptiness of existence affects people who are nevertheless very active. At the risk of sticking excessively to current news that may not speak to many people in a few years, let us return for a moment to the number of testimonies from people confined as a result of the COVID-19 crisis, forced to change their pace (but not always to lessen it, by the way) and becoming aware that the "normal" pace (the pace associated with the dominant conventional regime) is in fact not so "normal"[3]. This forced break has turned our daily life upside down in the sense that the conventions which govern

2 At the risk of repeating ourselves, we refer the reader who would like to know more about these myths to the book entitled *Le territoire entre liberté et durabilité* (Buclet 2011a).

3 While very few months elapsed between the confinement period and the writing of this chapter (July 2020), several publications are dedicated to this issue. Beyond the scientific work on the effects of confinement on the psycho-physical health of individuals, some texts question the possible benefits of this decrease in pace and the questions that follow on the meaning to be given to our existence (see, for example, Devillairs (2020) or Khermimoun (2020)).

our behavior and our relationships with others have had to be put aside, replaced by rules imposed by the emergency. Of course, deconfinement tends toward a return to the original conventions but leaves traces both in everyday life (interacting without physical contact) and in thinking about the values to be put forward. In other words, it is possible that this unprecedented period may produce long-term effects on the evolution of the dominant conventional regime by making more individuals aware of the weight of certain reference points on their existence; the myths we have outlined here. What is certain is that a conventional regime has significant stability and the one in which we live seems formidable from this point of view, especially since it tends to extend to the whole planet.

Indeed, as Robert Boyer and André Orléan (1994) have shown, there are four main modalities for switching from one convention to another: collapse, invasion, translation (rather valid for technical conventions) and agreement. The question that must be asked, even though a continuation of the dominant conventional regime is leading us toward an eventual collapse (scarcity of the resources necessary for a regime that needs more and more, climatic conditions unfavorable to many human activities, pressure on the environment exceeding the capacity of ecosystems to renew themselves, etc.), is that of our capacity to switch to a new conventional regime, with the characteristics of sustainability on the international political agenda for decades now.

5.2.2. *What might the characteristics be of a sustainable conventional regime?*

Changes are necessary in the face of the climate emergency and the degradation of a large number of ecosystems that are favorable to human societies. Everyone agrees on this, despite the persistence of a minority that denies the ecological emergency, a minority that is sometimes powerful when represented by the political authorities of leading countries (the United States and Brazil). The watchword in recent years has been transition. But what does transition mean?

Transition toward what? As highlighted in Chapter 1, according to the dominant players (national and international institutions, major economic companies, etc.), this transition would essentially involve technological progress, coupled with changes in behavior. Let us leave aside the technological solutions – since these depend closely on the direction taken and therefore on the conventional reference points that guide research – and focus first on the question of behavior. It seems to us that the latter is not addressed in any substantial way, for lack of questioning the foundations of our behaviors, i.e. the conventions that underlie them. More than a transition, it is a transformation that seems necessary to us. It is therefore in this direction that we propose to continue our reasoning.

As explained above, the framework of a conventional regime is made up of implicit or explicit reference points. The more dynamically stable a regime is, the more implicit these reference points are. Everyone refers to them without even thinking about them. Gradual changes over time, if consistent with conventional principles, reinforce the stability of the regime. It is therefore difficult to create disruption and thus transformation of a society without affecting the foundations of a regime, namely, its conventional reference points. A proactive attitude is necessary in order to move toward real change, in particular by proposing conventional principles capable of constituting the foundations of an alternative conventional regime in the longer term.

First of all, let us emphasize our desire to create an alternative within a democratic framework, not that this framework is ideal, but it does have a certain number of advantages, in particular that of preserving individual liberties as much as possible. From this perspective, our reflection is situated between two objectives: to move toward a society concerned with preserving individual liberties and both individual and collective initiatives; and to move toward a society that thinks of itself as a socio-ecosystem that is ecologically sustainable, while respecting the diversity of the contexts experienced by human communities.

5.2.2.1. *Increasing capabilities*

We believe that a link between an analysis through the metabolism of territories and a reflection on the principles to which human activities within communities should be anchored is likely to help reconcile these objectives. Thus, the reader will not be surprised that the first of the principles we wish to put forward is that of increasing capability, taken in its various meanings, as we saw in Chapter 4. More specifically, individual capability, at the basis of their freedom of action, collective capability, which makes it possible to generate collective dynamics (and which is generated by these same collective dynamics), and territorial capability, which anchors these collective capabilities in a precise context and in relation to the issues of ecological balance. This principle is based on a system of rights-goals, i.e.:

> a moral system in which respect for and violation of rights are included among the goals, integrated into the evaluation of situations and then applied to the choice of actions through links of cause and consequence. (Sen 1978, p. 130)

The aim of the policies to be implemented is to develop rights, which makes it possible to go beyond the simple defense of negative rights[4], to develop the capacity of individuals, collectives and territorial communities.

In order to increase these capabilities, beyond education aimed at making everyone aware of respect for others and differences of opinion (though should this not be an educational objective in any democracy?), it is fundamental to set up intermediary institutions in order to make effective the legislation favorable to the promotion of capabilities, according to the context. In particular, mechanisms for evaluating the effective respect of rights seem necessary (De Munck 2006, p. 260).

---

4 As a reminder, therefore, we do not limit ourselves to prohibiting individuals from harming the freedom of others or the community.

5.2.2.2. *Increasing participatory democracy*

These devices, both deliberative and informative, show the importance of another principle, that of increasing participatory democracy. The formal rights to be defended and enforced are deliberated upon and the results actually obtained are evaluated repeatedly within these same mechanisms. Citizen participation in these mechanisms is the best way to ensure that the decisions taken are adhered to in terms of the capabilities that need to be developed.

Above all, however, creating a society based on increased participation in decision-making processes, and in bodies that evaluate the impact of measures taken, is part of the need to reconcile freedom and collective constraints in the face of the challenges of sustainable development. If the trade-off to be made between the tendency of individuals to give priority to their own interests, and that of taking account of others, is made by those who will be concerned in the first place by the measures taken, the chances of adherence are all the greater. More precisely, we consider that procedures for evaluating the measures taken can be carried out with the help of territorial metabolism and that this monitoring would benefit from being carried out by the stakeholders of the territory concerned. The continuous dialogue between the development of territorial projects and the evaluation of their effects, in terms of the circulation of flows and thus of positive and negative impacts not only on the territory but also on the rest of the world, remains, in our opinion, a means of implementing a continuous process of participation in the future of a territory.

5.2.2.3. *Priority to the proximity principle*

This is only made possible on a certain scale, the scale likely to correspond to that of the feeling of belonging to a community of destiny. This seems indispensable to us for two reasons:

– as Alexis de Tocqueville has shown, in relation to the functioning of democracy in New England in the 19th century, it tends to weaken with the distance between citizens and the places not only where decisions are made, but also where decisions are executed. “Administrative centralization is only likely to irritate the peoples who submit to it, because it tends to constantly diminish the spirit of the city among them", he declares (Tocqueville 1835, p. 150). In contrast to centralization – both geographical and organizational proximity between citizens – administration and decision-making centers can only contribute to this "spirit of the city", which facilitates adherence of the greatest number to decisions that are sometimes to the detriment of their individual interests, and to the benefit of the community;

– the development of monitoring of flows generated by a human community, if it is possible to envisage it on the scale of entire countries (as we saw in Chapter 2), does not allow us to be sufficiently precise to establish a link between decision-making and the changes that this implies on the circulation of flows. For this, as we have shown, a metabolic representation not only of what circulates, but also of how and why it circulates, seems necessary and can only be done at the scale of living areas with territorial coherence.

The community of destiny is all the more cohesive because it is based on a plurality of proximities, as identified by the school of proximity (Filippi *et al.* 2018). Following in the footsteps of Bernard Pecqueur and Jean-Benoît Zimmermann (2004), we identify three types of proximity, all of which have a role to play in the promotion of a conventional regime that is an alternative to the one that dominates:

– Geographic proximity, which we saw with the spring 2020 lockdown, to what extent it depends on mobility capacities and means. One of the consequences of this geographical proximity is that interactions are less dependent on sophisticated technical devices whose malfunctions can be a source of vulnerability. Above all, however, being geographically close means a condition that favors – while not sufficient in itself – exchanges between individuals and their feeling of belonging to the same reality. Of course, geographical proximity does not only have virtues, especially if it is subject to and generates conflicts over the use of space and/or local resources (Torre 2018); and it is precisely the crossover between this geographical proximity, and the growth of a participatory democracy, that proves to be necessary to avoid too much proximity.

– Organizational proximity, which "concerns interactions between stakeholders (individual and collective), within (or between) organizations (and formal institutions) based on a common cognitive framework" (Colletis *et al.* 1999, p. 27). Participating in the same organization, or in related organizations pursuing compatible goals, promotes membership in the same community, not necessarily territorial in this case. In our view, and given the need to switch from one conventional regime to a more sustainable one, organizational proximity only makes sense if it reinforces the other two forms of proximity.

– Institutional proximity, in our view, plays, a fundamental role in the capacity to move toward a sustainable regime. This proximity rests "on the adherence of stakeholders to a common space of representations and rules of action guiding collective behavior. This institutional proximity, more or less strong, influences the compatibility of the different modes of coordination between stakeholders, and, consequently, the emergence of localized productive coordination regularities" (Colletis *et al.* 1999, p. 27). In other words, this proximity is linked to the sharing of conventions, be they interactional conventions (rules of action) or conventional reference points (which guide collective, as well as individual, behavior).

In other words, promoting different forms of proximity seems a good way to get human communities to adhere to the outline of an alternative conventional regime, while improving the capacity of individuals to adhere to the decisions that need to be made in the face of major ecological challenges. It also raises the question of economies of scale, another conventional benchmark, firmly established within the dominant conventional regime.

Fostering proximity, for all the reasons mentioned above, should make us aware of other aspects, including the meaning of what we do. Indeed, following the example of citizens who lose the meaning of the city according to Alexis de Tocqueville, André Gorz (1988) has shown us how work has lost its meaning for a large number of workers and employees. This loss of meaning is linked to distance at various levels: distance and lack of communication between the one who conceives what should be produced and the one who, through his or her work, participates in production; distance between the worker and the one who will have the utility of what is produced; and distance between the cognitive capacities of the worker and what he or she is asked to do. The incessant pursuit, considered indispensable, of economies of scale has been the main driving force behind these different forms of distancing. We can note that this has also affected the agricultural world. We find this in affluent countries where decisions about a number of operations (for example, when to treat crops with phytosanitary products) are made by the companies that produce the inputs and no longer by the farmers. It is also the case in poor countries, as in South America, with farms of hundreds of thousands of hectares that employ farm workers who are disconnected from the issues and meaning of what they do as simple performers of elementary tasks.

Furthermore, one of the challenges of the proximity principle is to allow each person to find meaning in what he or she does, not only because it concerns the territory in which he or she lives (geographical proximity), but also because it can be in line with the values in which he or she recognizes himself or herself (institutional proximity).

5.2.2.4. *Knowledge sharing*

What we have just highlighted relates to the question of who owns the knowledge mobilized within a society. Since the market has been put forward as an unsurpassable and much more efficient mode of coordination, the logic of competition is one of its operational translations. In the dominant conventional regime, knowledge is at the basis of the capacity to get ahead of competitors. Knowledge becomes private property and must therefore be protected so that no one, other than the person who generated it (even when it was in fact generated largely by public money, but here we are touching less on the foundation of the regime than on the absurd excesses that accompany it) can derive a competitive advantage from it. Knowledge is privatized, protected with patents (recurrent injunctions including in public universities), in order to exclude other beneficiaries. The societal argument to defend this state of affairs is that it is the best way, the most effective way in any case, to incite stakeholders to innovate, since they will have a substantial return on investment in the event of successful innovation. The idea that individuals and groups would want to think about and invent knowledge for future innovations, and thus respond to needs considered to be of prime necessity, does not seem to come to mind for those defending this point of view, even though this is what many researchers do on a daily basis[5]. The dominant conventional regime has shaped thinking (as any long-term conventional regime does) in a sense that humans act only out of self-interest and never out of altruism (see Chapter 3). This is largely true precisely because it has been instilled in us that it is human nature to be like this, and we therefore tend to reproduce this state of affairs. Yet the rise of multiple initiatives seems to show that, despite the strength of the dominant conventional reference points, many stakeholders are striving to act out of other motivations, including when it comes to innovation[6]. Concern about the drift of the notion of intellectual property, particularly over living things (Shiva 2001), does not seem to move policy makers, however, even though we are touching on the very heart of the mastery of knowledge as an immoderate amount of power.

In order to move toward a sustainable conventional regime, likely to increase individual, collective and territorial capabilities, knowledge sharing, increasingly practiced in the world of *open access*, is an indispensable condition. Indeed, it is through knowledge that individuals or collectives can gain autonomy and the ability to act according to their own objectives. It is also the dissemination of knowledge that can allow for more informed participatory debates on the consequences of the decisions that may be taken. This dissemination also makes it possible to reduce the effect of cognitively closed "expert systems" on themselves (Giddens 1990), i.e. to ensure

---

5 It would indeed be quite bold to argue that researchers are creative for profit, especially in countries such as France where research salaries are not very attractive. The argument is in fact a decoy, because companies that appropriate the knowledge generated by the world of public research would no longer have a sufficient knowledge base if the research world applied the same policy as they do with regard to the intellectual protection of discoveries.

6 Social innovation (Klein and Harrisson 2006), certainly not motivated by monetary returns on investment on the part of a large number of stakeholders practicing it, has probably been one of the most dynamic fields of innovation for a number of years.

that knowledge about our daily lives, marked by sophisticated technologies, can be shared or even that there is a need to explain how these expert systems function in order to feed participatory debates. So, it is therefore thanks to this sharing of knowledge that participatory debates can emerge from greater institutional proximity, i.e. shared values and principles.

5.2.2.5. *The principle of reciprocity*

Sharing knowledge is one way to stimulate reciprocity, but it cannot be the only one. One of the problems of many territories in the world is the nature of their relationships with stakeholders, whose main interests are outside the territory, whether they are located in other territories or are not linked to any particular territory (a-spatialized stakeholders such as multinationals). Unfavorable power relations substantially reduce a territory's capability and autonomy when it comes to deciding on the nature of wealth-creating activities and the type of flows that should be favored from a systemic perspective. Moreover, strategic knowledge may remain external, and territorial stakeholders may remain confined to operations with little cognitive content and therefore with little added value (a situation typical of colonialist relations). The purpose of adopting a principle of reciprocity would be to ensure that exchanges between territories – or even between parts of the world at a country level – should be balanced from the point of view of power relations between stakeholders (one of the primary characteristics of reciprocity, according to Karl Polanyi). In other words, it is not so much (or not only) the balance of physical flows circulating between territories that matters, as that of immaterial flows, including the interplay of influence and power between territories.

As we can see, these various principles are linked and feed off one other. Taken as a whole, their adoption aims to better reflect individual aspirations in terms of freedom, while refuting economic and financial liberalism, which is the source of deep inequalities and a lack of capacity to react firmly to ecological emergencies – and how could it do so since it is at the origin of them? In past works, we have mobilized some of these principles to show their relevance in very diverse contexts (Buclet 2011c). We will not focus here on specific examples, but rather attempt to question the relationship between the framework of an alternative conventional regime and innovation as a driving force for transformation.

5.3. Toward innovative regional projects

In the same way that we advocate transformative rather than transitive thinking in order to build a sustainable future, we wish to emphasize the question of the relationship between innovation and novelty, rather than between innovation and the market, as has been the case too much in the past. Innovation has become an omnipresent injunction, but is resolutely and primarily focused on the abiliy of countries to renew their capacity for economic growth. Guideines for a socio-ecological transition advocated by national and international institutions are rarely an exception to this framework, with the idea that a country which succeeds in developing the green technologies of the future will have a decisive competitive advantage. We can even, quite legitimately, suspect that a large number of economic stakeholders have no interest in tackling the problem at its source, given the increasing turnover in pollution control and treatment activities[7].

Without denying the need for technologies dedicated to solving the problems generated by the dominant conventional regime, it seems to

7 "Green industries are a fast-growing sector, with increasingly significant economic weight. Indeed, in terms of employment, they represented in France, in 2011, nearly 455,600 full-time equivalent employees (MEDDE, 2013), or +6.7% compared to 2010. A very significant increase compared to other economic sectors: +1.5%". Available at: https://www.ademe.fr/expertises/produire-autrement/production-industrielle-services/dossier/quoi-parle-t/eco-entreprises-filieres-vertes (accessed 22 July, 2020). These green sectors are: water and sanitation, recycling and waste recovery, renewable energies, and energy efficiency. We can also add emerging sectors, such as projects for carbon landfill, or green chemistry and biosourced products. Other sources confirm this progression: "between 2004 and 2015, the production of eco-activities (activities producing goods and services with the aim of protecting the environment or the sustainable management of resources) increased by an average of 4.2% per year compared to 2.3% for the economy as a whole". Available at: https://www.ecologique-solidaire.gouv.fr/sites/default/files/CEDD%20-%20Ref%20040.pdf, (accessed 22 July, 2020).

us necessary to resituate the search for technologies conducive to a reduction in ecological impact, in line with the conventional principles on which to rely in the future.

5.3.1. *What innovation?*

In the Green Paper on Innovation, written by the European Commission in 1995, "innovation is taken as synonymous with the successful production, assimilation and exploitation of novelty in the economic and social fields"[8]. We prefer Jean-Marc Fontan's proposal that innovation is "the process that takes place between the emergence of a new idea and the integration of its use in everyday life" (Fontan 2008). There is thus not only the emergence of a new idea, but also, and above all, its integration into the practices of a human community. This means that a good idea, as well as its development in a technical, organizational or institutional form, depends on its assimilation within the conventional regime in force. This is one of the points emphasized in the extensive, mostly Dutch, literature of the school of socio-technical transitions, of which Frank Geels is the leader (Geels 2002). Innovations are said to be generated either by stakeholders firmly rooted in the socio-technical regime, or by stakeholders who fit into niches. While the future of the former is fairly easily assured, niche innovations are more likely to be rejected because they are likely to be incompatible with the socio-technical regime. "The challenge of constructing the social use of a novelty therefore consists of making it legitimate and useful 'socially'" (Fontan 2006). This usefulness and social legitimacy therefore depends mainly on the conventional reference points in force. In the dominant conventional regime, an innovation can emerge and find its place if, for example, it promotes new gains in terms of economies of scale and thus a reduction in costs. This same innovation may be rejected in an alternative conventional regime based on the principles seen in this chapter, if this innovation contributes to various forms of remoteness, as well as to a reduction in territorial capability, both because of increased environmental

---

8 Available at: https://europa.eu/documents/comm/green_papers/pdf/com95_688_fr.pdf (accessed 22 July, 2020).

pressure and a loss of autonomy of the stakeholders in the territory in terms of production (economies of scale frequently lead to an increase in the concentration of productive activities).

The embedding of novelty in a social context is made possible by the legitimacy of this novelty from the point of view of the reference points of this context. According to Jean-Marc Fontan, it is a question of integrating novelty into a territorial fabric, because "without territory, there is no sociality, and without sociality, there is no territory!" (Fontan 2006). According to Fontan, however, these novelties are of two types: ordinary innovations (new individual behaviors that are ecologically preferable to previous behaviors, no longer using disposable bags, for example) and innovations with a "social innovation" dimension, i.e. capable of mobilizing stakeholders around a common social purpose (new forms of social networks that are potentially much more mobilizing to defend a cause, for example). Ordinary innovations, Jean-Marc Fontan tells us:

> as free atoms, can hardly be drivers of change. Linked with each other, put into shape, their transformative potential becomes more important. (Fontan 2006)

In this sense, local initiatives at the origin of even ordinary innovations, if they are linked with other local initiatives, are likely to give weight to the conventional reference points they underpin.

5.3.2. *Local initiatives at the heart of the alternative regime*

What do certain projects of Living Lab (Janin and Pecqueur 2016) and Fab Lab (see in particular the Usinette[9] project, whose specificities we described; Buclet 2015c), the Slow Movement (which brings together Slow Food but also Slow Cities), the TBS movement in India for water management by local communities, the Grameen Danone joint venture in Bangladesh, the *Città dell'Altra Economia*

---

9 Available at: https://usinette.org/actualites/article/au-confins-du-reel (accessed 22 July, 2020).

(City of the Other Economy) in Rome or the various associations promoting a solidarity economy have in common[10]? These are *a priori* very diverse realities and, above all, of very varied scale. They can range from very small initiatives (such as participatory housing, for example) to more regional initiatives (e.g. the Halwar district in Rajasthan for water management by and for local communities, about 6,500 km², roughly the equivalent of a French department, or a slightly larger area than Palestine), to international networks of local units, as in the case of the Slow Food movement present in 160 countries. Behind these very diverse realities and scales of action lie many of the principles introduced in this chapter. Whether explained in charters and other programmatic documents, or just observable through the action carried out by these organizations, capability (often translated as autonomy), scale of action in line with where people live, institutional proximity, reciprocity, and participation in decision-making processes are almost always present, as is knowledge sharing. With diverse modalities, taking into account various geographical, cultural and political contexts, a multitude of local initiatives gives substance to this aspiration for a different society. If the environmental issue is not systematically highlighted, in reality it very quickly becomes the counterpart to the development of initiatives, as long as they remain faithful to the original principles. In the same way, the concern for reciprocity and increased capacity is translated, most of the time, into actions in favor of the most disadvantaged. For these principles, as we have shown, are interrelated. They form a system. They cannot be chosen *à la carte*, according to the desire of the moment.

In view of these considerations, we can understand, having already established this finding on many occasions, that the reader is perplexed and, above all, skeptical in the face of what may appear to be localism. Indeed, it is a question of territory, of proximity and of renouncing economies of scale. All of this is very small in relation to the dominant conventional reference points that push us to grow by

10 For the reader interested in details about these local initiatives, these realities (and a few others) have already been described and analyzed elsewhere (Buclet 2011a, 2011c, 2015c).

going ever further, ever faster and producing ever more to ensure the well-being of humanity. Above all, we are well aware that power is very unevenly distributed across the planet and that this is a macroeconomic problem, clarified perfectly by economists such as Piketty (2013) or Saez and Zucman (2016). The dramatic increase in inequalities in capital ownership has worsened since the 1980s, making it difficult to speak of degrowth or sobriety when a large part of the world's population is trying to survive. We have been careful not to confront the question head-on because we believe that talking about degrowth is tantamount to validating the way of thinking of the dominant conventional regime. For in any regime, we measure and act essentially according to the conventions in force. To measure economic growth, even to criticize it, is to accept being locked into the dominant economic system and its conventional reference points. It is in this sense that we have evoked the creation of wealth in a form that is not exclusively monetary and, above all, mostly unmeasurable by money[11].

These spectacular monetary inequalities also have important consequences in terms of ecological impact. Chancel and Piketty (2015) show us that the richest 10% of the world are responsible for 45% of global emissions, while the poorest 50% are responsible for less than 13% of emissions. The richest 1% of the planet, they also tell us, emits 200 times more greenhouse gas emissions than the world's poorest populations. Above all, these inequalities are growing in parallel with the increase in monetary inequalities and pose the problem of political regulation of this phenomenon on a global scale.

5.3.3. *Is there a rise of an alternative regime?*

While we fully share the importance of these issues, as well as the proposals of these economists to stem the problems of inequality, we do not think that this will be enough in the face of climate change. Indeed, the proposals of Lucas Chancel and Thomas Piketty, however essential they may seem to us, do not imply a change in the

11 Economists are trying to do so, perpetuating the founding myths at the origin of our understanding of the world.

conventional regime which we have seen, whether or not there are strong inequalities, as the source of major ecological imbalances. Moreover, we do not see what would be more realistic between imagining a global regulation, or at least in the main countries, that would go against the most powerful and wealthy, and a rise in power of an alternative carried by a myriad of local initiatives that manage to develop smoothly with these same powerful and wealthy.

There is still the remark about the size of these initiatives, which is different from their generalization. Indeed, the issue of scaling up small-scale tested solutions is omnipresent in the discourse, as an absolute necessity for "getting down to business". However, a generalization of the ways of doing things, as practiced by these local initiatives with alternative values, may very well involve a multiplication of these initiatives rather than a change of scale. It is not that a change in the scale of initiatives is to be avoided, but it is essential that this change of scale is not detrimental to the values carried by the projects and is therefore in line with the principles set out to promote a sustainable society. The fear of a change in value in the event of a change of scale, and also of generalization, is not unfounded. The drift is just around the corner, as Julie Guthman (2004) has shown, in relation to the rise of organic agriculture in California, even though its share has only increased from 1% in the 1990s to 2% of agricultural land in 2002. If the pioneers of organic agriculture are faithful to their initial values, it goes without saying that the development of these practices can only be achieved through the conversion of conventional farmers. This state of affairs can only change the very practices of organic agriculture, unless we accompany farmers in conversion and make them aware of new principles. However, in reality, at least in California (though we can say the same in Europe with the European Union's organic label) the food industry is going to weigh in so that the entire organic sector takes into account its own commercial logic. To do so, this influential sector will push for a relaxation of organic specifications to better meet its requirements, to the point of bringing organic practices closer to those of conventional agriculture (Guthman 2004, p. 307). Julie Guthman also shows that producers who have switched to organic farming continue to grow most of their produce conventionally.

We could say, from the theory of socio-technical transitions, that organic, as a niche, has been absorbed by the dominant regime, with modification of some of its characteristics. The pioneers, or organic purists, then find themselves in competition not with conventional agriculture, but with more competitive organic products, although insufficiently respectful of ecological balances and social considerations. For consumers, with the exception of the most militant, there is no difference between these two types of organic and they do not see why they should pay more for organic when there are cheaper alternatives. In the same way, national and/or European organic subsidies do not make the distinction, as long as the specifications of the official organic label are respected.

More recently, the same questions have been raised about the growth of short food supply chains (Chiffoleau 2017). The movement, initially led by committed individuals for whom short food supply chains are part of a social project (Chiffoleau 2017, p. 131), has for several years benefited from the benevolence of local authorities, and even from an incentive to develop[12]. Younès Chiffoleau illustrates the polarization between these stakeholders, who offer an alternative, and the majority of agricultural organizations for whom the short circuit is above all a way to diversify distribution, because a clientele is attached to it[13]. In addition, the traditional agricultural world feels threatened by stakeholders who are not from the agricultural sector and who have other objectives. Younès Chiffoleau clearly shows the tensions between the two visions, then the arrival in the field of stakeholders acting according to the reference points of the dominant conventional regime. These are moments of questioning regarding the coexistence of different types of agriculture and the ability of a certain number of stakeholders to preserve transformative values in society.

The relatively effective response to these drifts and a reabsorption of the alternative by the conventional lie in the constitution of networks of multi-local stakeholders. In order not to compromise with the principles defended by a certain number of stakeholders, networks

12 In France, the regions have been responsible for the development of short circuits since 2012 (Chiffoleau 2017, p. 135).

13 In other words, this is just one market segment among others.

are being constituted and consolidated, in particular thanks to the Internet. Italy is particularly active in the food sector. In addition to the Slow Food movement, which is a fine international reality in the agri-food sector, other networks have been set up, such as the *Economia Solidale* (Solidarity Economy) network in 2002, based on the *Gruppi d'Acquisto Solidale* (GAS) (Solidarity Purchasing Groups or SPGs) that preexisted in the Italy of the 1990s[14]. This network links the grassroots associative structures that join together with these SPGs to form local networks and, in some regions, regional Solidarity Economy Networks (RES). All this is coordinated by the *Rete Italiana per l'Economia Solidale*, which became an association in January 2020[15]. This network is not only involved in the food sector, of course, but also in the fields of energy, insurance, especially car insurance, solidarity purchasing, education, culture, alternative travel, etc. Several members of the network are working to disseminate what they consider to be good practice or projects with political implications. Following the example of this Italian network, the multiplication of alternative stakeholders elsewhere in the world is significant. The recent rise in various European countries of complementary local currencies[16] is another important sign of this vivacity and the link with political issues. We cannot yet know what this actually represents, nor if this growth will not stop once the entire population sensitive to these values and principles has been mobilized. It is certain that many of these initiatives are based on the enthusiasm and activism of individuals and collectives who wish to propose an alternative political project. Carrying such an ambition is not without rest and real breathlessness in some cases. In particular, the dynamics around a certain number of complementary currencies, where they had emerged earlier, are running out of steam. For example, in Totnes in the United Kingdom, the "homeland" of cities in transition, the local currency has been gradually abandoned, without it really being clear whether this means that the dynamic is maturing or tending to run out of steam.

14 http://economiasolidale.net/content/chi-siamo (accessed 23 July, 2020).

15 This clearly shows the need for networking, rather than scaling up, at a pace that preserves the original principles.

16 Available at: https://base.socioeco.org/docs/_img_pdf_panorama_des_msc_en_europe-4.pdf (accessed 23 July, 2020).

We also see, through these initiatives and their attempt to become more generalized, that creativity is being promoted and leads to innovation in many areas. These innovations are at an organizational level and at the level of the link between values that should be emphasized, rather than at the technological level. However, as the rise of agro-ecology – and, more generally, biomass – and its uses show, technological innovations have their place. These innovations simply accompany a social project that is different from the current society and therefore have equally different characteristics. The search for efficiency and optimization is not ignored, but the proposed solutions make it a second-order objective, behind meaningful objectives such as sobriety, reciprocity, autonomy (and therefore the simplicity of the techniques mobilized, so that the greatest number can appropriate them) or even user-friendliness. If we observe what makes an innovation successful from a different point of view, we necessarily give much less importance to projects based on heavy investment and highly complex technologies, which distance individuals from the transformation processes from a cognitive, organizational and geographical point of view.

5.3.4. *The key role of politics*

Therefore, the challenge of generalization is to build effective support for these initiatives and the involvement of political stakeholders, first and foremost the elected representatives of the territories. Following the example in France of PATs (*Projets alimentaires territoriaux*, or Territorial Food Projects) – institutionalized by Law No. 2014-1170 of October 13, 2014 and known as the Law on the Future of Agriculture, Food and Forestry – an alliance is needed between territorial institutions, locally-elected officials and their services, and local initiatives. Local elected representatives should become aware of their pivotal role in the face of the challenges of social transformation. They have legitimacy, real though not absolute power, and they should see value in these initiatives, in terms of territorial development. Any urban, territorial, or metropolitan project, whatever it may be, should first and foremost be the consequence of a deliberative debate on the principles to be adopted and defended. Every territory should also provide itself with the means to monitor the evolution of a project, not only in terms of quantified objective measurements, but also in terms of changes in the modes of action and interactions within the territory and the evaluation of these changes in light of the principles adopted.

This also presupposes an awareness at the level of regional and national (or even European) public authorities that their role must be to accompany territorial dynamics from the moment they state their political project for development and that this project appears to be effectively transformative in the sense that we have defined it. What cements a nation, it should be remembered, is supposed to be found in a Constitution, that is, an agreement on a certain number of major principles that cannot be derogated from. Elected officials at the national level should remember that it is their responsibility to discuss the main guiding principles for the years and decades to come, and not just to pass laws based on current events, as has unfortunately been the case all too often in recent years. If so many citizens turn away from politics, we argue that one of the main causes is the inability of the political class to give meaning to society and to project itself according to objectives that are not only quantifiable but also desirable, in terms of life choices.

Finally, this presupposes that national and international institutions manage to adopt common rules to reallocate the remarkable financial resources that circulate around the world toward meaningful objectives, particularly ecological and social imbalances. But that, as Thomas Piketty has shown, is a different story altogether.

5.4. Conclusion

This chapter allowed us to go beyond what constitutes the core of territorial ecology and to project ourselves toward the means of building a sustainable future. Many researchers in this field are sensitive to this societal issue and do not work in research just for speculative pleasure. Therefore, it seemed important for us to make the link between what territorial ecology proposes and the political stakes to be faced. Here, we open the debate as to the relevance of putting forward alternative principles to the dominant conventional reference points, and of linking these principles with a territorial diagnostic approach inspired by territorial metabolism as we have proposed it. The transformation is today carried out by marginal stakeholders, given the dominant conventional regime. It is up to us as citizens, through our vote and political commitment, to make our representatives understand that a transformative project makes more sense than policies in support of an ecological transition that merely continues along the same dead-end path, because it is steered by those who have no interest in seeing the world really change.

Conclusion

With this book, I wanted to provide an overview of territorial ecology, both in terms of methods and issues, that was as complete as possible. The reader specialized in the elaboration of metabolisms, establishing the flows that circulate in a quantitatively precise way, will undoubtedly be a little disappointed. These methods have already proved their worth and it seemed to me that the methodological stakes were at another level: that of linking these important flow accounting tools with the other methodological issues facing territorial ecology.

For this reason, after recalling where and how territorial ecology is situated, both at the scientific and institutional levels, and after a short reminder of the main ways of mobilizing the concept of metabolism, I wished to use the last three chapters of this book to highlight a methodological proposal, in order to couple the scientific stakes with the operational issues that arise, both from the point of view of ecological urgency and from the economic and social point of view. The approach that consists of analyzing the territory as a system made up of wealth-creating activities is, in my view, particularly rich, in that it allows us to reconcile the issues of quantifying the footprint of human activities with the issues of understanding how and why human activities are structured and generate certain flows rather than others. In particular, I have been able to emphasize the importance that needs to be given to the whole intangible part of a territory's metabolism, whether it be know-how, territorial identity, heritage, power games or forms of coordination. Of course, this gives us a hybrid method that is necessarily less rigorous, since it is a question of representing quantifiable phenomena and others that can only be qualified. It depends on the objective we set ourselves. If the objective is operational, i.e. if we consider that territorial ecology can help territorial stakeholders in the elaboration phase of diagnosis, prospective and project monitoring, too great a precision in the quantitative measurement of flows seems less strategic than the capacity to make the dynamics of a territorial system explicit.

It is in this sense that I have insisted on the relationship between the metabolism of a territory and the capacities of territorial stakeholders to collectively take their destiny into their own hands and evolve in the face of events, be they climatic, sanitary or linked to economic relations with the rest of the world. A territorial diagnosis based on the analysis of territorial potential, particularly on its relationship with its environment and future developments, now seems to be an essential exercise in the face of so-called socio-ecological transition issues. The concept of territorial capability is interesting in this respect, combined with the notion of socio-ecological resources, as well as that of resilience. Although difficult to quantify, territorial capacity has the merit of questioning in a complex way:

– the way a territory operates, identified as a socio-ecosystem;

– the potential available to react or evolve according to events;

– how the players transform this potential into wealth while also preserving the potential for the future.

From an operational point of view, we consider that local authorities, or any other stakeholder concerned by territorial dynamics, would benefit from taking advantage of the proposed method in order to organize collective and deliberative exercises for the elaboration of their metabolism. The aim of these exercises would be to elaborate a diagnosis of the dynamics of the territorial system, from which to dialogue on a territorial project. But what I have tried to show in the last chapter is that a territorial project requires a reflection on the meaning of the project, understood from a political point of view. Faced with the challenges of sustainability, we can schematize two attitudes (in addition to the attitude that consists of denying human responsibility in the face of climate change):

1) change behaviors through economic incentives, empowering individuals through raising awareness and reducing the impact of our activities through technological and organizational innovation, in order to achieve ambitious climate goals while preserving the dominant conventional regime;

2) transforming society by adhering to a revisited system of values, adopting, through deliberation, principles of action that break away from the dominant conventional reference points, giving a different meaning to the future. I believe, moreover, that the term "transformation" is more appropriate to this second attitude than the transitional objective put forward by the institutions.

I have clearly taken sides in this debate, based on the analysis of the changes that I have been conducting over the past decade or so. It is quite logical that the dominant players do not want profound change, since their domination is based on the current regime. It is equally logical that the changes should emerge from stakeholders on the margins of the regime, since they are more likely to not recognize themselves in the values and objectives enacted. This is why I wanted to insist, like researchers in social innovation, on local initiatives carried out by stakeholders who are little recognized in the value system of the dominant conventional regime. Unless there is a collapse, which some people are calling for, without really knowing what this would mean, I am still betting on the alternative of the rise of modes of action and coordination that will break away from the dominant economic system. Perhaps all of this will not be enough, but that proposals complementary to our own are welcome, especially at the national and international levels. It would be a delusion to believe that global solutions could tip us towards a sustainable trajectory, given the impressive resistance to change on the part of the dominant players. We can only hope that such measures will not hinder the essential challenge of a conventional regime change.

Finally, my wish is to have convinced the reader of the relevance of the methodological contributions of territorial ecology, as well as of the contributions in terms of reflections on what makes it possible to understand a territorial dynamic and thus elaborate a territorial project, taking into account the socio-ecological stakes. I also hope to sensitize the territorial stakeholders to the question of the project of society, which must be in phase with the project of territory. The necessary transformation cannot do without a political debate on what is expected in the future and not just on what we would like to avoid in the future.

References

Adger, W.N. (2006). Vulnerability. *Global Environmental Change*, 16, 268–281.

Allenby, B., Richards, D., Frosch, R. (1994). *The Greening of Industrial Ecosystems: Overview and Perspectives.* National Academy Press, Washington, 259.

Ariès, P. (2005). *Décroissance ou barbarie.* Golias, Villeurbanne.

Artus, P. (2011). Une croissance durable? In *A la recherche de la nouvelle croissance*, Lorenzi, J.-H. (ed.). Presses Universitaires de France, Paris.

Ayres, R.U., Holmberg, J., Andersson, B. (2001). Materials and the global environment: Waste mining in the 21st century. *MRS Bull*, 26(6), 477–480.

Bairoch, P. (1993). *Economics and World History: Myths and Paradoxes.* University of Chicago Press, Chicago.

Barles, S. (2004). Regards d'écologues et de géochimistes sur la ville et la sociétéindustrielle, 1953–2003. In *Ville et durabilité : questions pour la recherche, questions à la recherche*, Barles, S., Traisnel, J.-P. (eds). Research Report carried out on behalf of the French Ministry of Research (ACI Ville). Laboratoire TMU(UMR CNRS AUS 7136), Champs-sur-Marne, June 2004, 175.

Barles, S. (2005). *L'invention des déchets urbains en France 1790–1990.* Editions Champ Wallon, Seyssel.

Barles, S. (2007). Mesurer la performance écologique des villes et des territoires : le métabolisme de Paris et de l'Ile-de-France. Final Report on behalf of the city of Paris, January 2007, 98.

Barles, S. (2009). Urban metabolism of Paris and its region. *Journal of Industrial Ecology*, 13(6), 898–913.

Barles, S. (2010). Ecologies urbaine, industrielle et territoriale. In *Ecologies Urbaines*, Coutard, O., Lévy, J.-P. (eds). Economica/Anthropos, Paris, 61–83.

Barles, S. (2014). L'écologie territoriale et les enjeux de la dématérialisation des sociétés : l'apport de l'analyse des flux de matières. *Développement durable et territoires*, 5(1).

Barles, S., Buclet, N., Billen, G. (2011). L'écologie territoriale : du métabolisme des sociétés à la gouvernance des flux d'énergie et de matières. *CIST2011 – Fonder les sciences du territoire*. Collège international des sciences du territoire (CIST), Paris, November 2011, 16–22.

Barles, S., Cerceau, J., Debuisson, M. (2015). Aussois et la Maurienne : une approche par les trajectoires socio-écologiques, XIXème–XXIème siècle. In *Essai d'écologie territoriale : l'exemple d'Aussois en Savoie*, Buclet, N. (ed.). CNRS Editions, 47–103.

Baudrillard, J. (1970). *La société de consommation*. Denoël, Paris.

Bentham, J. (1888). *Principes de législation et d'économie politique*. Guillaumin & cie, Paris (republished in 2007 by Elibron Classics).

Berlin, I. (1969). *Four Essays on Liberty*. Oxford University Press, Oxford.

Berthoud, G. and Kilani, M. (1984). L'exemple de l'agriculture à temps partiel en milieu alpin. *Uni-Lausanne*, 38, 1–15.

Bevione, M. (2021). L'analyse de flux de matière pour caractériser les enjeux socio-écologiques d'une filière à différentes échelles géographiques: le cas de la filière laitière bovine française et de la fabrication du fromage AOP Beaufort dans la vallée de la Maurienne. Thesis, Université Grenoble Alpes.

Billen, G., Toussaint, F., Peeters, P., Sapir, M., Steenhout, A., Vanderborght, J.-P. (1983). *L'écosystème belgique : essai d'écologie industrielle*. Centre de Recherche et d'Information Socio-Politique (CRISP), Brussels.

Billen, G., Barles, S., Garnier, J., Rouillard, J., Benoit, P. (2009). The food-print of Paris: Long-term reconstruction of the nitrogen flows imported into the city from its rural hinterland. *Regional Environmental Change*, 9(1), 13–24.

Billen, G., Garnier, J., Barles, S. (2012a). History of the urban environmental imprint: Introduction to a multidisciplinary approach to the long-term relationships between Western cities and their hinterland. *Regional Environmental Change*, 12(2), 249–252.

Billen, G., Chatzimpiros, P., Barles, S., Garnier, J. (2012b). Grain and meat to feed Paris: The changing balance between cereal cultivation and animal farming in the Paris Basin (18th–21st centuries). *Regional Environmental Change*, 12(2), 325–335.

Billen, G., Garnier, J., Lassaletta, L. (2013). Le cycle de l'azote : moteur du fonctionnement des écosystèmes. In *Vers une nouvelle synthèse écologique : de l'écologie scientifique au développement durable*, Meerts, P. (ed.). CIVA, ULB, Brussels, 124–147.

Boltanski, L. and Chiappello, E. (1999). *Le nouvel esprit du capitalisme*. Gallimard, Paris.

Boltanski, L. and Thévenot, L. (1991). *De la justification : les économies de la grandeur*. Gallimard, Paris.

Bonaudo, T., Billen, G., Garnier, J., Barataud, F., Bognon, S., Marty, P., Dupré, D. (2015). Le système agro-alimentaire : un découplage progressif de la production et de la consommation. In *Essai d'écologie territoriale : l'exemple d'Aussois en Savoie*, Buclet, N. (ed.). CNRS Editions, 161–180.

Boyer, R. and Orléan, A. (1994). Persistance et changement des conventions. Deux modèles simples et quelques illustrations. In *Analyse économique des conventions*, Orléan, A. (ed.). PUF, Paris, 219–247.

Bringezu, S. and Moriguchi, Y. (2002). Material flow analysis. In *A Handbook of Industrial Ecology*, Ayres, R.U., Ayres, L.W. (eds). Edward Elgar Publishing, Cheltenham, Northampton, 79–90.

Brullot, S. (2009). Mise en œuvre de projets territoriaux d'écologie industrielle en France : vers un outil méthodologique d'aide à la décision. Thesis, Université de Technologie de Troyes, Troyes.

Buclet, N. (2011a). *Le territoire entre liberté et durabilité*. Presses Universitaires de France, Paris.

Buclet, N. (2011b). *Ecologie industrielle et territoriale : stratégies locales pour un développement durable*. Presses Universitaires du Septentrion, Villeneuve d'Ascq.

Buclet, N. (2011c). Territoire, innovation et développement durable : l'émergence d'un nouveau régime conventionnel ? *Revue d'economie regionale et urbaine*, 5, 911–940.

Buclet, N. (2015a). Ecologie industrielle et économie circulaire : définitions et principes. In *Economie circulaire et écosystèmes portuaires*, Alix, Y., Cerceau, J., Mat, N. (eds). Editions EMS, Caen, 27–41.

Buclet, N. (ed.) (2015b). *Essai d'écologie territoriale : l'exemple d'Aussois en Savoie*. CNRS Editions, Paris.

Buclet, N. (2015c). Le territoire créateur de ressources : une illustration autour du cas d'Usinette. *Flux*, 1, 47–57.

Buclet, N. and Cerceau, J. (2019). Interactions et rétroactions entre dimensions matérielle et immatérielle de systèmes communs de ressources spatialisés, une lecture par l'écologie territoriale. *Développement durable et territoires*, 10(1) [Online]. Available: https://doi.org/10.4000/developpementdurable.13467.

Buclet, N. and Donsimoni, M. (2018). L'écologie territoriale : où comment resituer l'économie au-delà de la sphère monétaire. In *Renouveler la géographie économique*, Pecqueur, B., Talandier, M. (eds). Economica, Paris, 188–203.

Buclet, N. and Lazarevic, D. (2015). Principles for sustainability: The need to shift to a sustainable conventional regime. *Environment, Development and Sustainability*, 17(1), 83–100.

Buclet, N., Cerceau, J., Debuisson, M., Maillefert, M., Métereau, R., Pech, P., Talandier, M. (2015). Création de richesses et réponses aux besoins de la population d'Aussois. In *Essai d'écologie territoriale : l'exemple d'Aussois en Savoie*, Buclet, N. (ed.). CNRS Editions, 105–160.

Caillé, A. (2009). *Théorie anti-utilitariste de l'action : fragments d'une sociologie générale*. La Découverte, Paris.

Chancel, L. and Piketty, T. (2015). *Carbon and Inequality: From Kyoto to Paris. Trends in the Global Inequality of Carbon Emissions (1998–2013) & Prospects for an Equitable Adaptation Fund*. School of Economics, Paris [Online]. Available: http://www.ledevoir.com/documents/pdf/chancelpiketty2015.pdf [Accessed 20 June 2020].

Char, R. (1983). Recherche de la base et du sommet. *Œuvres complètes*. Gallimard, Pléïade, Paris.

Chatzimpiros, P. (2011). Les empreintes environnementales de l'approvisionnement alimentaire : Paris, ses viandes et lait, XIXème–XXIème siècles. Thesis, Université Paris-Est, Paris.

Chiffoleau, Y. (2017). Dynamique des identités collectives dans le changement d'échelle des circuits courts alimentaires. *Revue française de socio-économie*, 18, 123–141.

Clastres, P. (1972). *Chronique des Indiens Guayaki*. Plon, Paris.

Clastres, P. (1974). *La société contre l'Etat*. Editions de Minuit, Paris.

Colletis, G., Gilly, J.-P., Leroux, I., Pecqueur, B., Perrat, J., Rychen, F., Zimmermann, J.-B. (1999). Construction territoriale et dynamiques productives. *Sciences de la société*, 48, 25–46.

Commission Européenne (2012). L'innovation au service d'une croissance durable : une bioéconomie pour l'Europe. Communication de la Commissionau Parlement Européen, au Conseil, au Comité Economique et Social Européen et au Comité des Régions, COM (2012) 60, Brussels, 13 February 2012.

Crouzat, E., Mouchet, M., Turkelboom, F., Byczek, C., Meermands, J., Berger, F., Verkerk, P.J., Lavorel, S. (2015). Assessing bundles of ecosystem services from regional to landscape scale: Insights from the French Alps. *Journal of Applied Ecology*, 1145–1155 [Online]. Available: https://doi.org/10.1111/1365-2664.12502 [Accessed 18 June 2020].

Crutzen, P.J. (2006). The anthropocene. In *Earth System Science in the Anthropocene*, Ehlers, E., Krafft, T. (eds). Springer, Berlin, 13–18.

Dardot, P. and Laval, C. (2015). *Commun : essai sur la révolution au XXIème siècle*, La Découverte, Paris.

Dauphiné, A. and Provitolo, D. (2007). La résilience : un concept pour la gestion des risques. *Annales de Géographie*, 2(654), 115–125.

De Munck, J. (2006). Vers un nouveau paradigme du droit. In *L'économie des conventions : méthodes et résultats, tome 1: débats*, Eymard-Duvernay, F. (ed.). La Découverte, Paris, 249–262.

Descola, P. (2005). *Par-delà nature et culture*. NRF Editions Gallimard, Paris.

Devillairs, L. (2020). Méditations pascalienne sur le confinement. *Etudes*, 5, 45–54.

Díaz, S., Unai, P., Stenseke, M., Martín-López, B., Watson, R.T., Molnár, Z., Hill, R., Chan, K.M.A., Baste, I.A., Brauman, K.A., Polasky, S., Church, A., Lonsdale, M., Larigauderie, A., Leadley, P.W., van Oudenhoven, A.P.E., van der Plaat, F., Schröter, M., Lavorel, S., Aumeeruddy-Thomas, Y., Bukvareva, E., Davies, K., Demissew, S., Erpul, G., Failler, P., Guerra, C.A., Hewitt, C.L., Keune, H., Lindley, S., Shirayama, Y. (2018). Assessing nature's contributions to people. *Science*, 359(6373), 270–272.

Druckman, A. and Jackson, T. (2006). Understanding households as drivers of carbon emissions. In *Taking Stocks of Industrial Ecology*, Clift, R., Druckman, A. (eds). Springer, Berlin, 181–204.

Dubois, S. (2017). Mutations du marché international du lait : une erreur d'appréciation européenne ? *La revue internationale et strategique*, 105, 46–55.

Duret, B. (2006). Premiers retours d'expériences en écologie industrielle : études de cas en Europe et en Amérique du Nord. Synthèse et perspectives. *Cahiers de la Chaire d'Ecologie Industrielle*, Université de Technologie de Troyes, Troyes.

Duret, B., Mat, N., Bonard, A., Dastrevigne, E., Lafragette, A. (2007). Ecologie territoriale : une aide à la définition d'une politique énergétique. Comprendre l'économie physique du territoire. *Les Annales de la Recherche Urbaine*, 103, 73–78.

Duvigneaud, P. (1974). *La synthèse écologique : populations, communautés, écosystèmes, biosphère, noosphère*. Doin, Paris.

Ehrenfeld, J.R. (2000). Industrial ecology: Paradigm shift or normal science. *American Behavioral Scientist*, 44(2), 229–244.

Ehrenfeld, J.R. and Chertow, M.R. (2002). Industrial symbiosis: The legacy of Kalundborg. In *A Handbook of Industrial Ecology*, Ayres, R.U., Ayres, L.W. (eds). Edward Elgar Publishing, Cheltenham, 334–348.

Erb, K.-H., Haberl, H., Jepsen, M.R., Kuemmerle, T., Lindner, M., Müller, D., Verburg, P.H., Reenberg, A. (2013). A conceptual framework for analysing and measuring land use intensity. *Current Opinion in Environmental Sustainability*, 5, 464–470.

Erb, K.-H., Fetzel, T., Haberl, H., Kastner, T., Kroisleitner, C., Lauk, C., Niedertscheider, M., Plutzar, C. (2016). *Social Ecology: Society-Nature Relations across Time and Space*, Haberl, H., Fischer-Kowalski, M., Krausmann, F., Winiwarter, V. (eds). Springer, Berlin, 5, 93–124.

Erkman, S. (2004). *Vers une écologie industrielle : comment mettre en pratique le développement durable dans une société hyper-industrielle*, 2nd edition. Editions Charles Léopold Mayer, Paris.

Esposito, R. (2019). *Communauté, immunité, biopolitique : repenser les termes de la politique*. Editions Mimesis, Sesto San Giovanni.

Evans, P. (2002). Collective capabilities, culture, and Amartya Sen's development as freedom. *Studies in Comparative International Development*, 37(2), 54–60.

Faure, M. (1999). Un produit agricole "affiné" en objet culturel. *Terrain*, 33 [Online]. Available: http://terrain.revues.org/2703 [Accessed 10 June 2020].

Filippi, M., Wallet, F., Polge, E. (2018). L'école de la proximité : naissance et évolution d'une communauté de connaissance. *Revue d'economie regionale et urbaine*, 5–6, 939–966.

Fischer-Kowalski, M. and Weisz, H. (2016). The archipelago of social ecology and the island of the Vienna school. In *Social Ecology: society-Nature Relations across Time and Space*, Haberl, H., Fischer-Kowalski, M., Krausmann, F., Winiwarter, V. (eds), Springer, Berlin, 3–28.

Fivos-Sargentis, G., Ioannidis, R., Karakatsanis, G., Sigourou, S., Lagaros, N., Koutsoyiannis, D. (2019). The development of the Athens water supply system and inferences for optimizing the scale of water infrastructures. *Sustainability*, 11(9) [Online]. Available: https://doi.org/10.3390/su11092657 [Accessed 13 April 2020].

Fontaine, G. (2016). Susciter l'émergence de communs comme outils du développement durable. *Actes de la 12ème conference internationale de l'AFD sur le developpement : communs et developpement*, 1–2 December 2016, 11.

Fontan, J.-M. (2008). Développement territorial et innovation sociale : l'apport polanyien. *Interventions économiques* [Online]. Available: https://doi.org/10.4000/interventionseconomiques.369.

Frosch, R. and Gallopoulos, N. (1989). Strategies for manufacturing. *Scientific American*, 261, 144–152.

Gallaud, D. (2020). Proximité organisée "subie" et effets négatifs des proximités. *Revue d'economie regionale et urbaine*, 1, 113–128.

Geels, F. (2002). Technological transitions as evolutionary reconfiguration processes: A multi-level perspective and a case-study. *Research Policy*, 31, 1257–1274.

Georgescu-Roegen, N. (1971). *The Entropy Law and the Economic Process*. Harvard University Press, Cambridge.

Giddens, A. (1990). *Les conséquences de la modernité*. L'Harmattan, Paris.

Gimpel, J. (1975). *La révolution industrielle du Moyen-Âge*. Editions du Seuil, Paris.

Godard, O. (1990). Environnement, modes de coordination et systèmes de légitimité : analyse de la catégorie de patrimoine naturel. *Revue Economique*, 41(2), 215–242.

Godbout, J. (1992). *L'esprit du don*. La Découverte, Paris.

Godbout, J. (2007). *Ce qui circule entre nous : donner, recevoir, rendre*. Le Seuil, Paris.

Gorz, A. (1988). *Métamorphose du travail : critique de la raison économique*. Le Seuil, Paris.

Greene, D., Kahn, J., Gibson, R. (1999). Fuel economy rebound effects for US household vehicles. *Energy*, 20(3), 1–29.

Grosinger, J. (2021). Analysis of the interaction between nature and society: The integration of ecosystem services in wealth-creating activities in the valley of Maurienne. Thesis, Université Grenoble Alpes, Grenoble.

Guthman, J. (2004). The trouble with "Organic Lite" in California: A rejoinder to the "conventionalisation" debate. *Sociologia Ruralis*, 44(3), 301–316.

Harchaoui, S. and Chatzimpiros, P. (2017). Reconstructing production efficiency, land use and trade for livestock systems in historical perspective. The case of France, 1961–2010. *Land Use Policy*, 67, 378–386.

Hardin, G. (1968). The tragedy of the commons. *Science*, 162, 1243–1248.

Herbelin, A. (2018). Ecologie territoriale et trajectoires de transitions : le cas du Rhône-Médian. Thesis, Université Grenoble Alpes, Grenoble.

Héritier, S. (2007). Energie et environnement : l'exploitation des sables bitumineux en Alberta. *Mappemonde*, 87, 3 [Online]. Available: https://mappemonde-archive.mgm.fr/num15/articles/art07304.pdf.

Herring, H. (2006). Energy efficiency – A critical review. *Energy*, 31(1), 10–20.

Hopkins, R. (2010). *Manuel de transition : de la dépendance au pétrole à la résilience locale*. Editions Ecosociété, Montreal.

Ibrahim, S.S. (2006). From individual to collective capabilities: The capability approach as a conceptual framework for self-help. *Journal of Human Development*, 7(3), 397–416.

Ingold, A. (2008). Les sociétés d'irrigation : bien commun et action collective. *Entreprises et Histoire*, 50, 19–35.

Janin, C. and Pecqueur, B. (2016) Les Living Labs : remise en question des processus de mise en marché et de politique publique. *Revue canadienne des sciences regionales*, 40(1).

Janssen, M.A. (2006). Resilience, vulnerability, and adaptation: A cross-cutting theme of the International Human Dimensions Programme on Global Environmental Change. *Global Environmental Change*, 16, 237–239.

Keane, D. (2012). Medieval London and its supply hinterlands. *Regional Environmental Changes*, 12(2), 263–281.

Kébir, L. and Crévoisier, O. (2004). Dynamique des ressources et milieux innovateurs. In *Ressources naturelles et culturelles, milieux et développement local*, Maillat, D., Camagni, R., Mattéaccioli, A. (eds). Edes-GREMI, Neuchatel, 261–290.

Keynes, J.M. (1936). *The General Theory of Employment, Interest, and Money*. Cambridge University Press, Cambridge.

Khermimoun, J. (2020). Sortir de l'impasse [Online]. Available: http://cerii.fr/produit/sortir-de-limpasse-lorsque-le-coronavirus-suscite-espoir-et-optimisme/. ⟨hal-02567764⟩.

Kilani, M. (2012). *Anthropologie : du local au global*. Armand Colin, Paris.

Kim, E. (2013). Les transitions énergétiques urbaines du XIXème au XXIème siècle : de la biomasse aux combustibles fossiles et fissiles à Paris (France). Thesis, Université Panthéon-Sorbonne – Paris I, Paris.

Klein, J.-L. and Harrisson, D. (eds) (2006). *L'innovation sociale : émergence et effets sur la transformation des sociétés*. Presses de l'Université du Québec, Quebec.

Kolm, S.-C. (1984). *La bonne économie : la réciprocité générale*. PUF, Paris.

Krausmann, F., Weisz, H., Eisenmenger, N. (2016). Transitions in sociometabolic regimes throughout human history. In *Social Ecology: Society-Nature Relations across Time and Space*, Haberl, H., Fischer-Kowalski, M., Krausmann, F., Winiwarter, V. (eds), Springer, Berlin, 5, 63–92.

Lafaye, C. and Thévenot, L. (1993). Une justification écologique ? Conflits dans l'aménagement de la nature. *Revue française de sociologie*, 34(4), 495–524.

Laval, C. (2007). *L'homme économique : essai sur les racines du néolibéralisme*. Gallimard, Paris.

Le Page, C. (2017). Simulation multi-agents interactive : engager des populations locales dans la modélisation des socio-écosystèmes pour stimuler l'apprentissage social, HDR, Université Pierre et Marie Curie, Sorbonne Universités, Paris.

Lessmann, O. and Rauschmayer, F. (eds) (2014). *The Capability Approach and Sustainability*. Routledge, Taylor & Francis Group, London.

Lévi-Strauss, C. (1958). *Anthropologie structurale*. Plon, Paris.

Lifset, R. (1997). A metaphor, a field and a journal. *Journal of Industrial Ecology*, 1(1), 1–3.

Livet, P. and Thévenot, L. (1994). Les catégories de l'action collective. In *Analyse économique des conventions*, Orléan, A. (ed.). PUF, Paris, 139–167.

Loubet, F., Dissart, J.-C., Lallau, B. (2011). Approche par les capacités à l'évaluation du développement territorial. *Revue d'economie regionale et urbaine*, 4, 681–703.

Magnaghi, A. (2012). *La biorégion urbaine : petit traité sur le territoire bien commun*. Eterotopia France, Rhizome, Paris.

Malinowski, B. (1922). *Les Argonautes du Pacifique occidental*. Gallimard, Paris.

Mathieu, J. (2000). *Storia delle Alpi 1500–1900 : Ambiente, sviluppo e società*. Edizioni Casagrande, Bellinzona.

Mauroy, H. (2011). La fable des abeilles de Bernard Mandeville : l'exploitation de son prochain comme fondement de la civilisation. *Revue europeenne des sciences sociales*, 49(1), 83–110.

Meadows, D.H., Meadows, D.L., Randers, J., Behrens III, W.W. (1972). *The Limits to Growth*. Potomac Associates, Washington DC.

Mélo, A. (2015). Fruitières comtoises : de l'association de voisins au district agro-industriel : une histoire montagnarde ? *Journal of Alpine Research*, 103(1) [Online]. Available: http://rga.revues.org/2785; DOI :10.4000/rga.2785 [Accessed 27 July 2017].

Ménégoz, M., Valla, E., Jourdain, N., Blanchet, J., Beaumet, J., Wilhelm, B., Gallée, H., Fettweis, X., Morin, S., Anquetin, S. (2020). Contrasting seasonal changes in total and intense precipitation in the European Alps from 1903 to 2010. *Hydrology and Earth System Sciences Discussions* [Online]. Available: https://doi.org/10.5194/hess-2019-690 [Accessed 18 June 2020].

Meybeck, M., Lestel, L., Bonté, P., Moilleron, R., Colin, J.-L., Rousselot, O., Hervé, D., de Pontevès, C., Grosbois, C., Thévenot, D.R. (2007). Historical perspective of heavy metals contamination (Cd, Cr, Cu, Hg, Pb, Zn) in the Seine River basin (France) following a DPSIR approach (1950–2005). *Science of the Total Environment*, 375(1–3), 204–231.

Milanesi, J. (2010). Éthique et évaluation monétaire de l'environnement : la nature est-elle soluble dans l'utilité ? *VertigO*, 10(2) [Online]. Available: https://doi.org/10.4000/vertigo.10050.

Ministère de l'Agriculture et de l'Alimentation (2016). *Une stratégie bioéconomie pour la France : enjeux et vision* [Online]. Available: file:///C:/Users/bucletn/AppData/Local/Temp/bioeconomie-fr-bd.pdf.

Ministère de l'Environnement (2007). Rapport du groupe 6. Modes de développement écologique favorables à l'emploi et à la compétitivité, Grenelle de l'environnement. Working group reports, Paris.

Mitchell, K., Agle, B., Wood, D. (1997). Toward a theory of stakeholder identification and salience: Defining a principle of who and what really counts. *Academy of Management Review*, 22(4), 853–886.

Nahon, D. (2008). *L'épuisement de la terre, l'enjeu du XXIème siècle*. Odile Jacob, Paris.

Nelson, D.R., Adger, W.N., Brown, K. (2007). Adaptation to environmental change: Contributions of a resilience framework. *Annual Review of Environment and Resources*, 32, 395–419.

Odum, E. (1963). *Ecologie : un lien entre les sciences naturelles et les sciences humaines*. Doin, Paris.

Odum, E. (1975). *Ecologie : un lien entre les sciences naturelles et les sciences humaines*, 2nd edition. Doin, Paris.

Ostrom, E. (2009). A general framework for analyzing sustainability of socio-ecological systems. *Sciences*, 325(5939), 419–422.

Pecqueur, B. and Zimmermann, J.-B. (2004). *Economie de proximités*. Hermès Sciences Publications, Lavoisier, Paris.

Perry, N. (1990). *Symbiosis: Nature in Partnership*. Blanford Press, London.

Piketty, T. (2013). *Le capital au XXIème siècle*. Le Seuil, Paris.

Polanyi, K. (1944). *La grande transformation*. Gallimard, Paris.

Polanyi, K. (2008). *Essais*. Le Seuil, Paris.

Rabourdin, S. (2012). *Replanter les consciences. Une refondation de la relation Homme/Nature*. Yves Michel. Gap.

Rahnema, M. (2003). *Quand la misère chasse la pauvreté*. Actes Sud, Paris.

Rauschmayer, F., Bauler, T., Schäpke, N. (2015). Towards a thick understanding of sustainability transitions – linking transition management, capabilities and social practices. *Ecological Economics*, 109, 211–221.

République du Canton de Genève (2005). *Ecologie industrielle à Genève : premiers résultats et perspective*. Service Cantonal de Gestion des Déchets [Online]. Available: http://www.icast.org/fichiers/Ecosite/EI_Metabolisme_GE.pdf.

Rist, G. (2001) *Le développement : histoire d'une croyance occidentale*, 2nd edition. Presses de Sciences Po, Paris.

Rivier, C. (2019). Circulation de flux matériels et enjeux socio-écologiques de la production agricole à l'échelle territoriale : le cas de la production du fromage AOP Beaufort dans la vallée de la Maurienne. Report, ENSAT, Auzeville-Tolosane.

Sachs, I. (1974). Environnement et styles de développement. *Annales*, 29(3), 553–570.

Saez, E. and Zucman, G. (2016). Wealth inequality in the United States since 1913: Evidence from capitalized income tax data. *The Quarterly Journal of Economics*, 131(2), 519–578.

Sahlins, M. (1973). *Age de pierre âge d'abondance*. Gallimard, Paris.

Sen, A. (1987). *Ethique et économie et autres essais*. PUF, Paris.

Sen, A. (1999). *Un nouveau modèle économique : développement, justice, liberté*. Odile Jacob, Paris.

Sen, A. (2014). The ends and means of sustainability. In *The Capability Approach and Sustainability*, Lessmann, O. and Rauschmayer, F. (eds). Routledge, Taylor & Francis Group, London, 6–20.

Singleton, M. (2010). *Histoires d'eaux africaines*. Editions Académia-Bruylant, Louvain la Neuve.

Shiva, V. (2001). *Patents, Myths and Reality*. Penguins Books India, New Dehli.

Smit, B. and Wandel, J. (2006). Adaptation, adaptive capacity and vulnerability. *Global Environmental Change*, 16, 282–292.

Smith, A. (1776). *The Wealth of Nations*. W. Strahan and T. Cadell, London.

Stergiouli, M. and Hadjibiros, K. (2012). The growing water imprint of Athens (Greece) throughout history. *Regional Environmental Change*, 12(2), 337–345.

Sugden, R. (1989). Spontaneous order. *Journal of Economic Perspectives*, 3(4), 85–97.

Swaney, D.P., Santoro, R.L., Howarth, R.W., Hong, B., Donaghy, K.P. (2012). Historical changes in the food and water supply systems of the New York City metropolitan area. *Regional Environmental Change*, 12(2), 363–380.

Taguieff, P.-A. (2001). *Résister aubougisme : démocratie forte contre mondialisation techno-marchande*. Editions des mille et une nuits, Paris.

de Tocqueville, A. (1835). *De la démocratie en Amérique I*. Gallimard, Paris.

Torre, A. (2018). Développement territorial et relations de proximité. *Revue d'economie regionale et urbaine*, 5–6, 1043–1075.

Van Parijs, P. (1991). Les deux écologismes. In *La pensée écologiste : essai d'inventaire à l'usage de ceux qui la pratiquent comme de ceux qui la craignent*, De Roose, F., Van Parijs, P. (eds). De Boeck Université, Brussels.

Villar, C. and David, M. (2014). La résilience, un outil pour les territoires. *Séminaire IT-GO Rosko*, Roscoff, 22–23 May.

Vitousek, P.M., Ehrlich, P.R., Ehrlich, A.H., Matson, P.A. (1986). Human appropriation of the products of photosynthesis. *BioScience*, 36, 368–373.

Wolman, A. (1965). The metabolism of cities. *Scientific American*, 213(3), 179–190.

Zaccaï, E. (2002). *Le développement durable : dynamique et constitution d'un projet*. Presses Interuniversitaires Européennes, Brussels.

Zaoual, H. (2006). *Management situé et développement local*. Horizon Pluriel, Rabat.

Index

K, L, M

N, P

R, S

T

U, V, W

Other titles from

iSTE

in

Innovation, Entrepreneurship and Management

2021

BOBILLIER CHAUMON Marc-Eric
Digital Transformations in the Challenge of Activity and Work: Understanding and Supporting Technological Changes
(Technological Changes and Human Resources Set – Volume 3)

2020

ACH Yves-Alain, RMADI-SAÏD Sandra
Financial Information and Brand Value: Reflections, Challenges and Limitations

ANDREOSSO-O'CALLAGHAN Bernadette, DZEVER Sam, JAUSSAUD Jacques, TAYLOR Robert
Sustainable Development and Energy Transition in Europe and Asia
(Innovation and Technology Set – Volume 9)

BEN SLIMANE Sonia, M'HENNI Hatem
Entrepreneurship and Development: Realities and Future Prospects
(Smart Innovation Set – Volume 30)

CHOUTEAU Marianne, FOREST Joëlle, NGUYEN Céline
Innovation for Society: The P.S.I. Approach
(Smart Innovation Set – Volume 28)

CORON Clotilde
Quantifying Human Resources: Uses and Analysis
(Technological Changes and Human Resources Set – Volume 2)

CORON Clotilde, GILBERT Patrick
Technological Change
(Technological Changes and Human Resources Set – Volume 1)

CERDIN Jean-Luc, PERETTI Jean-Marie
The Success of Apprenticeships: Views of Stakeholders on Training and Learning
(Human Resources Management Set – Volume 3)

DELCHET-COCHET Karen
Circular Economy: From Waste Reduction to Value Creation
(Economic Growth Set – Volume 2)

DIDAY Edwin, GUAN Rong, SAPORTA Gilbert, WANG Huiwen
Advances in Data Science
(Big Data, Artificial Intelligence and Data Analysis Set – Volume 4)

DOS SANTOS PAULINO Victor
Innovation Trends in the Space Industry
(Smart Innovation Set – Volume 25)

GASMI Nacer
Corporate Innovation Strategies: Corporate Social Responsibility and Shared Value Creation
(Smart Innovation Set – Volume 33)

GOGLIN Christian
Emotions and Values in Equity Crowdfunding Investment Choices 1: Transdisciplinary Theoretical Approach

GUILHON Bernard
Venture Capital and the Financing of Innovation
(Innovation Between Risk and Reward Set – Volume 6)

LATOUCHE Pascal
Open Innovation: Human Set-up
(Innovation and Technology Set – Volume 10)

LIMA Marcos
Entrepreneurship and Innovation Education: Frameworks and Tools
(Smart Innovation Set – Volume 32)

MACHADO Carolina, DAVIM J. Paulo
Sustainable Management for Managers and Engineers

MAKRIDES Andreas, KARAGRIGORIOU Alex, SKIADAS Christos H.
Data Analysis and Applications 3: Computational, Classification, Financial, Statistical and Stochastic Methods
(Big Data, Artificial Intelligence and Data Analysis Set – Volume 5)
Data Analysis and Applications 4: Financial Data Analysis and Methods
(Big Data, Artificial Intelligence and Data Analysis Set – Volume 6)

MASSOTTE Pierre, CORSI Patrick
Complex Decision-Making in Economy and Finance

MEUNIER François-Xavier
Dual Innovation Systems: Concepts, Tools and Methods
(Smart Innovation Set – Volume 31)

MICHAUD Thomas
Science Fiction and Innovation Design (Innovation in Engineering and Technology Set – Volume 6)

MONINO Jean-Louis
Data Control: Major Challenge for the Digital Society
(Smart Innovation Set – Volume 29)

MORLAT Clément
Sustainable Productive System: Eco-development versus Sustainable Development
(Smart Innovation Set – Volume 26)

SAULAIS Pierre, ERMINE Jean-Louis
Knowledge Management in Innovative Companies 2: Understanding and Deploying a KM Plan within a Learning Organization
(Smart Innovation Set – Volume 27)

2019

AMENDOLA Mario, GAFFARD Jean-Luc
Disorder and Public Concern Around Globalization

BARBAROUX Pierre
Disruptive Technology and Defence Innovation Ecosystems
(Innovation in Engineering and Technology Set – Volume 5)

DOU Henri, JUILLET Alain, CLERC Philippe
Strategic Intelligence for the Future 1: A New Strategic and Operational Approach
Strategic Intelligence for the Future 2: A New Information Function Approach

FRIKHA Azza
Measurement in Marketing: Operationalization of Latent Constructs

FRIMOUSSE Soufyane
Innovation and Agility in the Digital Age
(Human Resources Management Set – Volume 2)

GAY Claudine, SZOSTAK Bérangère L.
Innovation and Creativity in SMEs: Challenges, Evolutions and Prospects
(Smart Innovation Set – Volume 21)

GORIA Stéphane, HUMBERT Pierre, ROUSSEL Benoît
Information, Knowledge and Agile Creativity
(Smart Innovation Set – Volume 22)

HELLER David
Investment Decision-making Using Optional Models
(Economic Growth Set – Volume 2)

HELLER David, DE CHADIRAC Sylvain, HALAOUI Lana, JOUVET Camille
The Emergence of Start-ups
(Economic Growth Set – Volume 1)

HÉRAUD Jean-Alain, KERR Fiona, BURGER-HELMCHEN Thierry
Creative Management of Complex Systems
(Smart Innovation Set – Volume 19)

LATOUCHE Pascal
Open Innovation: Corporate Incubator
(Innovation and Technology Set – Volume 7)

LEHMANN Paul-Jacques
The Future of the Euro Currency

LEIGNEL Jean-Louis, MÉNAGER Emmanuel, YABLONSKY Serge
Sustainable Enterprise Performance: A Comprehensive Evaluation Method

LIÈVRE Pascal, AUBRY Monique, GAREL Gilles
Management of Extreme Situations: From Polar Expeditions to Exploration-Oriented Organizations

MILLOT Michel
Embarrassment of Product Choices 2: Towards a Society of Well-being

N'GOALA Gilles, PEZ-PÉRARD Virginie, PRIM-ALLAZ Isabelle
Augmented Customer Strategy: CRM in the Digital Age

NIKOLOVA Blagovesta
The RRI Challenge: Responsibilization in a State of Tension with Market Regulation
(Innovation and Responsibility Set – Volume 3)

PELLEGRIN-BOUCHER Estelle, ROY Pierre
Innovation in the Cultural and Creative Industries
(Innovation and Technology Set – Volume 8)

PRIOLON Joël
Financial Markets for Commodities

QUINIOU Matthieu
Blockchain: The Advent of Disintermediation

RAVIX Joël-Thomas, DESCHAMPS Marc
Innovation and Industrial Policies
(Innovation between Risk and Reward Set – Volume 5)

ROGER Alain, VINOT Didier
Skills Management: New Applications, New Questions
(Human Resources Management Set – Volume 1)

SAULAIS Pierre, ERMINE Jean-Louis
Knowledge Management in Innovative Companies 1: Understanding and Deploying a KM Plan within a Learning Organization
(Smart Innovation Set – Volume 23)

SERVAJEAN-HILST Romaric
Co-innovation Dynamics: The Management of Client-Supplier Interactions for Open Innovation
(Smart Innovation Set – Volume 20)

SKIADAS Christos H., BOZEMAN James R.
Data Analysis and Applications 1: Clustering and Regression, Modeling-estimating, Forecasting and Data Mining
(Big Data, Artificial Intelligence and Data Analysis Set – Volume 2)
Data Analysis and Applications 2: Utilization of Results in Europe and Other Topics
(Big Data, Artificial Intelligence and Data Analysis Set – Volume 3)

UZUNIDIS Dimitri
Systemic Innovation: Entrepreneurial Strategies and Market Dynamics

VIGEZZI Michel
World Industrialization: Shared Inventions, Competitive Innovations and Social Dynamics
(Smart Innovation Set – Volume 24)

2018

BURKHARDT Kirsten
Private Equity Firms: Their Role in the Formation of Strategic Alliances

CALLENS Stéphane
Creative Globalization
(Smart Innovation Set – Volume 16)

CASADELLA Vanessa
Innovation Systems in Emerging Economies: MINT – Mexico, Indonesia, Nigeria, Turkey
(Smart Innovation Set – Volume 18)

CHOUTEAU Marianne, FOREST Joëlle, NGUYEN Céline
Science, Technology and Innovation Culture
(Innovation in Engineering and Technology Set – Volume 3)

CORLOSQUET-HABART Marine, JANSSEN Jacques
Big Data for Insurance Companies
(Big Data, Artificial Intelligence and Data Analysis Set – Volume 1)

CROS Françoise
Innovation and Society
(Smart Innovation Set – Volume 15)

DEBREF Romain
Environmental Innovation and Ecodesign: Certainties and Controversies
(Smart Innovation Set – Volume 17)

DOMINGUEZ Noémie
SME Internationalization Strategies: Innovation to Conquer New Markets

ERMINE Jean-Louis
Knowledge Management: The Creative Loop
(Innovation and Technology Set – Volume 5)

GILBERT Patrick, BOBADILLA Natalia, GASTALDI Lise,
LE BOULAIRE Martine, LELEBINA Olga
Innovation, Research and Development Management

IBRAHIMI Mohammed
Mergers & Acquisitions: Theory, Strategy, Finance

LEMAÎTRE Denis
Training Engineers for Innovation

LÉVY Aldo, BEN BOUHENI Faten, AMMI Chantal
Financial Management: USGAAP and IFRS Standards
(Innovation and Technology Set – Volume 6)

MILLOT Michel
Embarrassment of Product Choices 1: How to Consume Differently

PANSERA Mario, OWEN Richard
Innovation and Development: The Politics at the Bottom of the Pyramid
(Innovation and Responsibility Set – Volume 2)

RICHEZ Yves
Corporate Talent Detection and Development

SACHETTI Philippe, ZUPPINGER Thibaud
New Technologies and Branding
(Innovation and Technology Set – Volume 4)

SAMIER Henri
Intuition, Creativity, Innovation

TEMPLE Ludovic, COMPAORÉ SAWADOGO Eveline M.F.W.
Innovation Processes in Agro-Ecological Transitions in Developing Countries
(Innovation in Engineering and Technology Set – Volume 2)

UZUNIDIS Dimitri
Collective Innovation Processes: Principles and Practices
(Innovation in Engineering and Technology Set – Volume 4)

VAN HOOREBEKE Delphine
The Management of Living Beings or Emo-management

2017

AÏT-EL-HADJ Smaïl
The Ongoing Technological System
(Smart Innovation Set – Volume 11)

BAUDRY Marc, DUMONT Béatrice
Patents: Prompting or Restricting Innovation?
(Smart Innovation Set – Volume 12)

BÉRARD Céline, TEYSSIER Christine
Risk Management: Lever for SME Development and Stakeholder Value Creation

CHALENÇON Ludivine
Location Strategies and Value Creation of International Mergers and Acquisitions

CHAUVEL Danièle, BORZILLO Stefano
The Innovative Company: An Ill-defined Object
(Innovation between Risk and Reward Set – Volume 1)

CORSI Patrick
Going Past Limits To Growth

D'ANDRIA Aude, GABARRET Inés
Building 21st Century Entrepreneurship
(Innovation and Technology Set – Volume 2)

DAIDJ Nabyla
Cooperation, Coopetition and Innovation
(Innovation and Technology Set – Volume 3)

FERNEZ-WALCH Sandrine
The Multiple Facets of Innovation Project Management
(Innovation between Risk and Reward Set – Volume 4)

FOREST Joëlle
Creative Rationality and Innovation
(Smart Innovation Set – Volume 14)

GUILHON Bernard
Innovation and Production Ecosystems
(Innovation between Risk and Reward Set – Volume 2)

HAMMOUDI Abdelhakim, DAIDJ Nabyla
Game Theory Approach to Managerial Strategies and Value Creation
(Diverse and Global Perspectives on Value Creation Set – Volume 3)

LALLEMENT Rémi
Intellectual Property and Innovation Protection: New Practices and New Policy Issues
(Innovation between Risk and Reward Set – Volume 3)

LAPERCHE Blandine
Enterprise Knowledge Capital
(Smart Innovation Set – Volume 13)

LEBERT Didier, EL YOUNSI Hafida
International Specialization Dynamics
(Smart Innovation Set – Volume 9)

MAESSCHALCK Marc
Reflexive Governance for Research and Innovative Knowledge
(Responsible Research and Innovation Set – Volume 6)

MASSOTTE Pierre
Ethics in Social Networking and Business 1: Theory, Practice and Current Recommendations
Ethics in Social Networking and Business 2: The Future and Changing Paradigms

MASSOTTE Pierre, CORSI Patrick
Smart Decisions in Complex Systems

MEDINA Mercedes, HERRERO Mónica, URGELLÉS Alicia
Current and Emerging Issues in the Audiovisual Industry
(Diverse and Global Perspectives on Value Creation Set – Volume 1)

MICHAUD Thomas
Innovation, Between Science and Science Fiction
(Smart Innovation Set – Volume 10)

PELLÉ Sophie
Business, Innovation and Responsibility
(Responsible Research and Innovation Set – Volume 7)

SAVIGNAC Emmanuelle
The Gamification of Work: The Use of Games in the Workplace

SUGAHARA Satoshi, DAIDJ Nabyla, USHIO Sumitaka
Value Creation in Management Accounting and Strategic Management: An Integrated Approach
(Diverse and Global Perspectives on Value Creation Set –Volume 2)

UZUNIDIS Dimitri, SAULAIS Pierre
Innovation Engines: Entrepreneurs and Enterprises in a Turbulent World
(Innovation in Engineering and Technology Set – Volume 1)

2016

BARBAROUX Pierre, ATTOUR Amel, SCHENK Eric
Knowledge Management and Innovation
(Smart Innovation Set – Volume 6)

BEN BOUHENI Faten, AMMI Chantal, LEVY Aldo
Banking Governance, Performance And Risk-Taking: Conventional Banks Vs Islamic Banks

BOUTILLIER Sophie, CARRÉ Denis, LEVRATTO Nadine
Entrepreneurial Ecosystems (Smart Innovation Set – Volume 2)

BOUTILLIER Sophie, UZUNIDIS Dimitri
The Entrepreneur (Smart Innovation Set – Volume 8)

BOUVARD Patricia, SUZANNE Hervé
Collective Intelligence Development in Business

GALLAUD Delphine, LAPERCHE Blandine
Circular Economy, Industrial Ecology and Short Supply Chains
(Smart Innovation Set – Volume 4)

GUERRIER Claudine
Security and Privacy in the Digital Era
(Innovation and Technology Set – Volume 1)

MEGHOUAR Hicham
Corporate Takeover Targets

MONINO Jean-Louis, SEDKAOUI Soraya
Big Data, Open Data and Data Development
(Smart Innovation Set – Volume 3)

MOREL Laure, LE ROUX Serge
Fab Labs: Innovative User
(Smart Innovation Set – Volume 5)

PICARD Fabienne, TANGUY Corinne
Innovations and Techno-ecological Transition
(Smart Innovation Set – Volume 7)

2015

CASADELLA Vanessa, LIU Zeting, DIMITRI Uzunidis
Innovation Capabilities and Economic Development in Open Economies
(Smart Innovation Set – Volume 1)

CORSI Patrick, MORIN Dominique
Sequencing Apple's DNA

CORSI Patrick, NEAU Erwan
Innovation Capability Maturity Model

FAIVRE-TAVIGNOT Bénédicte
Social Business and Base of the Pyramid

GODÉ Cécile
Team Coordination in Extreme Environments

MAILLARD Pierre
Competitive Quality and Innovation

MASSOTTE Pierre, CORSI Patrick
Operationalizing Sustainability

MASSOTTE Pierre, CORSI Patrick
Sustainability Calling

2014

DUBÉ Jean, LEGROS Diègo
Spatial Econometrics Using Microdata

LESCA Humbert, LESCA Nicolas
Strategic Decisions and Weak Signals

2013

HABART-CORLOSQUET Marine, JANSSEN Jacques, MANCA Raimondo
VaR Methodology for Non-Gaussian Finance

2012

DAL PONT Jean-Pierre
Process Engineering and Industrial Management

MAILLARD Pierre
Competitive Quality Strategies

POMEROL Jean-Charles
Decision-Making and Action

SZYLAR Christian
UCITS Handbook

2011

LESCA Nicolas
Environmental Scanning and Sustainable Development

LESCA Nicolas, LESCA Humbert
Weak Signals for Strategic Intelligence: Anticipation Tool for Managers

MERCIER-LAURENT Eunika
Innovation Ecosystems

2010

SZYLAR Christian
Risk Management under UCITS III/IV

2009

COHEN Corine
Business Intelligence

ZANINETTI Jean-Marc
Sustainable Development in the USA

2008

CORSI Patrick, DULIEU Mike
The Marketing of Technology Intensive Products and Services

DZEVER Sam, JAUSSAUD Jacques, ANDREOSSO Bernadette
Evolving Corporate Structures and Cultures in Asia: Impact of Globalization

2007

AMMI Chantal
Global Consumer Behavior

2006

BOUGHZALA Imed, ERMINE Jean-Louis
Trends in Enterprise Knowledge Management

CORSI Patrick *et al.*
Innovation Engineering: the Power of Intangible Networks

Printed and bound by CPI Group (UK) Ltd, Croydon, CR0 4YY
15/09/2026
14962353-0002